Calculus with Complex Numbers

Calculus with Complex Numbers

John B. Reade

First published 2003
by Taylor & Francis
11 New Fetter Lane, London EC4P 4EE

Simultaneously published in the USA and Canada
by Taylor & Francis Inc.
29 West 35th Street, New York, NY 10001

Taylor & Francis is an imprint of the Taylor & Francis Group

Typeset in Times New Roman by
Newgen Imaging Systems (P) Ltd, Chennai, India
Printed and bound in Great Britain by
MPG Books Ltd, Bodmin

British Library Cataloguing in Publication Data
A catalogue record for this book is available from the British Library

Library of Congress Cataloging in Publication Data
A catalog record for this book has been requested

ISBN 0–415–30846–1 (hbk)
ISBN 0–415–30847–X (pbk)

Contents

Preface

This book is based on the premise that the learning curve is isomorphic to the historical curve. In other words, the learning order of events is the same as the historical order of events. For example, we learn arithmetic before we learn algebra. We learn how before we learn why.

Historically, calculus with real numbers came first, initiated by Newton and Leibnitz in the seventeenth century. Even though complex numbers had been known about from the time of Fibonacci in the thirteenth century, nobody thought of doing calculus with complex numbers until the nineteenth century. Here the pioneers were Cauchy and Riemann. Rigorous mathematics as we know it today did not come into existence until the twentieth century. It is important to observe that the nineteenth century mathematicians had the right theorems, even if they didn't always have the right proofs.

The learning process proceeds similarly. Real calculus comes first, followed by complex calculus. In both cases we learn by using calculus to solve problems. It is when we have seen what a piece of mathematics can do that we begin to ask whether it is rigorous. Practice always comes before theory.

The emphasis of this book therefore is on the applications of complex calculus, rather than on the foundations of the subject. A working knowledge of real calculus is assumed, also an acquaintance with complex numbers. A background not unlike that of an average mathematician in 1800. Equivalently, a British student just starting at university. The approach is to ask what happens if we try to do calculus with complex numbers instead of with real numbers. We find that parts are the same, whilst other parts are strikingly different. The most powerful result is the residue theorem for evaluating complex integrals. Students wishing to study the subject at a deeper level should not find that they have to unlearn anything presented here.

I would like to thank the mathematics students at Manchester University for sitting patiently through lectures on this material over the years. Also for their feedback (positive and negative) which has been invaluable. The book is respectfully dedicated to them.

John B. Reade
June 2002

Preface

Chapter 1

Complex numbers

1.1 The square root of minus one

Complex numbers originate from a desire to extract square roots of negative numbers. They were first taken seriously in the eighteenth century by mathematicians such as de Moivre, who proved the first theorem in the subject in 1722. Also Euler, who introduced the notation i for $\sqrt{-1}$, and who discovered the mysterious formula $e^{i\theta} = \cos\theta + i\sin\theta$ in 1748. And third Gauss, who was the first to prove the fundamental theorem of algebra concerning existence of roots of polynomial equations in 1799. The nineteenth century saw the construction of the first model for the complex numbers by Argand in 1806, later known as the Argand diagram, and more recently as the complex plane. Also the first attempts to do calculus with complex numbers by Cauchy in 1825. Complex numbers were first so called by Gauss in 1831. Previously they were known as imaginary numbers, or impossible numbers. It was not until the twentieth century that complex numbers found application to science and technology, particularly to electrical engineering and fluid dynamics.

If we want square roots of negative numbers it is enough to introduce $i = \sqrt{-1}$ since then, for example, $\sqrt{-2} = \sqrt{-1}\sqrt{2} = i\sqrt{2}$. Combining i with real numbers by addition and multiplication cannot produce anything more general than $x + iy$ where x, y are real. This is because the sum and product of any two numbers of this form are also of this form. For example,

$$
\begin{aligned}
(1+2i)+(3+4i) &= 4+6i,\\
(1+2i)(3+4i) &= 3+10i+8i^2\\
&= 3+10i-8\\
&= -5+10i.
\end{aligned}
$$

Subtraction produces nothing new since, for example,

$$(1+2i)-(3+4i) = -2-2i.$$

Neither does division since, for example,

$$\frac{1+2i}{3+4i} = \frac{(1+2i)(3-4i)}{(3+4i)(3-4i)} = \frac{3+2i-8i^2}{9-16i^2}$$
$$= \frac{3+2i+8}{9+16} = \frac{11+2i}{25} = \frac{11}{25} + \frac{2}{25}i.$$

The number $3-4i$ is called the *conjugate* of $3+4i$. For any $x+iy$ we have

$$(x+iy)(x-iy) = x^2 + y^2$$

so division can always be done except when $x = y = 0$, that is, when $x+iy = 0$.

It is also possible to extract square roots of numbers of the form $x+iy$ as numbers of the same form. For example, suppose

$$\sqrt{1+2i} = A + iB,$$

then we have

$$1+2i = (A+iB)^2 = A^2 + 2iAB - B^2.$$

So we require

$$A^2 - B^2 = 1,$$
$$AB = 1.$$

The second equation gives $B = 1/A$, which on substitution in the first equation gives

$$A^2 - \frac{1}{A^2} - 1 = 0.$$

Solving this quadratic equation in A^2 by the formula we obtain

$$A^2 = \frac{1 \pm \sqrt{5}}{2}.$$

For real A we must take

$$A^2 = \frac{\sqrt{5}+1}{2},$$

which gives

$$B^2 = \frac{1}{A^2} = \frac{2}{\sqrt{5}+1} = \frac{\sqrt{5}-1}{2}.$$

Hence we obtain

$$\sqrt{1+2i} = \sqrt{\frac{\sqrt{5}+1}{2}} + i\sqrt{\frac{\sqrt{5}-1}{2}}.$$

This last property of numbers of the form $x+iy$ represents a bonus over what might reasonably have been expected. Introducing square roots of negative real numbers is one thing. Creating a number system in which square roots can always be taken is asking rather more. But this is precisely what we have achieved. Existence of square roots means that quadratic equations can always be solved. We shall see shortly that much more is true, namely that polynomial equations of any degree can be solved with numbers of the form $x+iy$. This is the fundamental theorem of algebra (see Chapter 8).

1.2 Notation and terminology

If $i = \sqrt{-1}$, then numbers of the form $x + iy$ are called *complex* numbers. We write $z = x + iy$ and call x the *real* part of z which we abbreviate to $\operatorname{Re} z$, and y the *imaginary* part of z which we abbreviate to $\operatorname{Im} z$.

N.B. $\operatorname{Re} z$, $\operatorname{Im} z$ are both *real*.

For $z = x+iy$ we write (by definition) $\bar{z} = x - iy$, and call $\bar{z}$ the *conjugate* of z.

For $z = x + iy$ we write (by definition) $|z| = \sqrt{x^2 + y^2}$, and call $|z|$ the *modulus* of z.

For example, if $z = 3 + 4i$ we have $\operatorname{Re} z = 3$, $\operatorname{Im} z = 4$, $\bar{z} = 3 - 4i$, and

$$|z| = \sqrt{3^2 + 4^2} = \sqrt{25} = 5.$$

1.3 Properties of $\bar{z}$, $|z|$

We list the fundamental properties of $\bar{z}$, $|z|$.

1. $z\bar{z} = |z|^2$. To see this observe that if $z = x + iy$, then

$$z\bar{z} = (x+iy)(x-iy) = x^2 + y^2 = |z|^2.$$

2. $\operatorname{Re} z = (z+\bar{z})/2$, $\operatorname{Im} z = (z-\bar{z})/2i$. To see this observe that if $z = x+iy$, then

$$z + \bar{z} = (x+iy) + (x-iy) = 2x, \quad z - \bar{z} = (x+iy) - (x-iy) = 2iy.$$

3. $\overline{z+w} = \bar{z} + \bar{w}$. To see this observe that if $z = x+iy$, $w = u+iv$, then

$$z + w = (x+iy) + (u+iv) = (x+u) + i(y+v),$$

therefore we have

$$\overline{z+w} = (x+u) - i(y+v) = (x-iy) + (u-iv) = \bar{z} + \bar{w}.$$

4. $\overline{zw} = \bar{z}\,\bar{w}$. To see this observe that if $z = x + iy$, $w = u + iv$, then

$$zw = (x+iy)(u+iv) = (xu - yv) + i(xv + yu),$$
$$\bar{z}\,\bar{w} = (x-iy)(u-iv) = (xu - yv) - i(xv + yu).$$

5. $|zw| = |z|\,|w|$. We delay the proof of this property until Section 1.9.
6. $|z+w| \le |z| + |w|$. We delay the proof of this property until Section 1.11.

1.4 The Argand diagram

We obtain a geometric model for the complex numbers by representing the complex number $z = x + iy$ by the point (x, y) in the real plane with coordinates x and y.

Observe that the horizontal x-axis represents complex numbers $x + iy$ with $y = 0$, that is, the real numbers. We therefore call the horizontal axis the *real* axis. The vertical y-axis represents complex numbers $x + iy$ with $x = 0$, that is, numbers of the form iy where y is real. We call these numbers *pure imaginary*, and we call the vertical axis the *imaginary* axis. The origin O represents the number zero which is of course real (Figure 1.1).

1.5 Geometric interpretation of addition

If we have two complex numbers $z = x + iy$, $w = u + iv$, then their sum $z + w$ is given by

$$z + w = (x+u) + i(y+v)$$

and therefore appears on the Argand diagram as the vector sum of z and w.

The complex number $z + w$ is represented geometrically as the fourth vertex of the parallelogram formed by 0, z, w (see Figure 1.2). For example, $3 + 2i$ is the vector sum of 3 and $2i$ (see Figure 1.1).

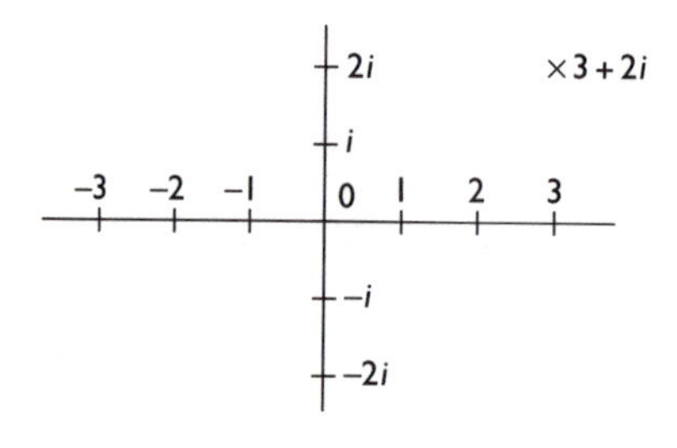

Figure 1.1

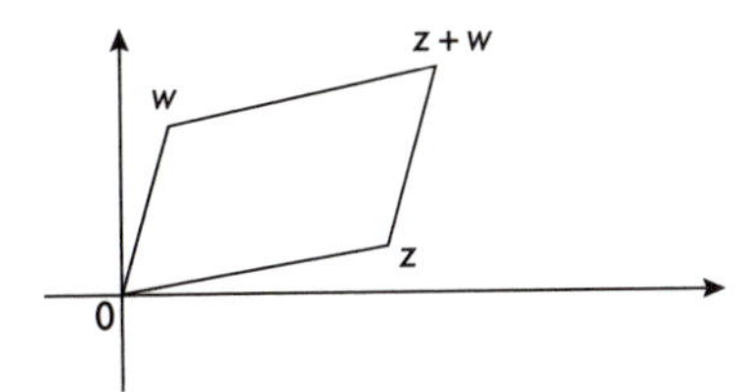

Figure 1.2

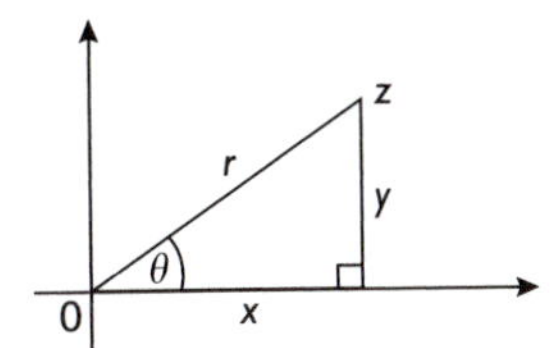

Figure 1.3

1.6 Polar form

An alternative representation of points in the plane is by polar coordinates r, θ. The coordinate r represents the distance of the point from the origin O. The coordinate θ represents the angle the line joining the point to O makes with the positive direction of the x-axis measured anticlockwise (see Figure 1.3). Suppose the complex number $z = x + iy$ on the Argand diagram has polar coordinates r, θ. We call r the *modulus* of z, and denote it by $|z|$. Pythagoras' theorem gives

$$|z| = \sqrt{x^2 + y^2}$$

consistent with the definition of $|z|$ given in Section 1.2.

We call θ the *argument* of z which we abbreviate to $\arg z$. A little trigonometry on Figure 1.3 gives

$$\theta = \tan^{-1}\frac{y}{x} = \sin^{-1}\frac{y}{r} = \cos^{-1}\frac{x}{r}.$$

Observe that whilst $|z|$ is single valued, $\arg z$ is *many* valued. This is because for any given value of θ we could take instead $\theta + 2\pi$ (in radians) and arrive at the same complex number z. For example, suppose $z = 1 + i$. Then $|z| = \sqrt{2}$, but $\arg z$ can be taken to be any of the values $\pi/4$, $5\pi/4$, $9\pi/4$, etc., also $-3\pi/4$, $-7\pi/4$, etc. Equivalently, $\arg z = \pi/4 + 2n\pi$ for any integer n.

We define the *principal value* (PV) of $\arg z$ to be that value of θ which satisfies $-\pi < \theta \leq \pi$. For example, the principal value of $\arg(1 + i)$ is $\pi/4$ (Figure 1.4).

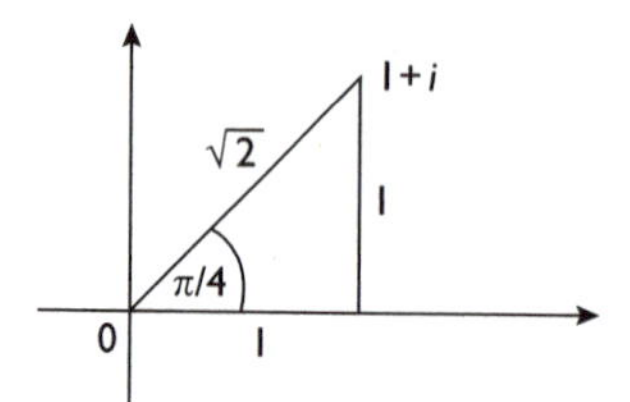

Figure 1.4

We write

$$\arg(1+i) = \pi/4 \text{ (PV)}.$$

For general $z = x + iy$ we have $\cos\theta = x/r$, $\sin\theta = y/r$ (see Figure 1.3). Therefore

$$\begin{aligned} z &= x + iy \\ &= r\cos\theta + ir\sin\theta \\ &= r(\cos\theta + i\sin\theta) \\ &= re^{i\theta}, \end{aligned}$$

since, by Taylor's theorem,

$$\begin{aligned} e^{i\theta} &= 1 + i\theta + \frac{(i\theta)^2}{2!} + \frac{(i\theta)^3}{3!} + \frac{(i\theta)^4}{4!} + \cdots \\ &= 1 + i\theta - \frac{\theta^2}{2!} - i\frac{\theta^3}{3!} + \frac{\theta^4}{4!} + \cdots \\ &= \left(1 - \frac{\theta^2}{2!} + \frac{\theta^4}{4!} - \cdots\right) + i\left(\theta - \frac{\theta^3}{3!} + \cdots\right) \\ &= \cos\theta + i\sin\theta. \end{aligned}$$

We call the formula

$$e^{i\theta} = \cos\theta + i\sin\theta$$

Euler's formula. We call the representation $z = re^{i\theta}$ the *polar form* for z. We call the representation $z = x + iy$ the *Cartesian form* for z. For example, $1 + i = \sqrt{2}e^{i\pi/4}$ (see Figure 1.4).

1.7 De Moivre's theorem

An immediate consequence of Euler's formula (see Section 1.6) is the result known as *de Moivre's* theorem, viz.,

$$(\cos\theta + i\sin\theta)^n = (e^{i\theta})^n = e^{in\theta} = \cos n\theta + i\sin n\theta.$$

Application 1 We can use de Moivre's theorem to obtain formulae for $\cos n\theta$, $\sin n\theta$ in terms of $\cos\theta$, $\sin\theta$. For example, we have

$$\begin{aligned}\cos 2\theta + i\sin 2\theta &= (C + iS)^2\\ &= C^2 + 2iCS + i^2S^2\\ &= (C^2 - S^2) + 2iCS,\end{aligned}$$

where $C = \cos\theta$, $S = \sin\theta$. Equating real and imaginary parts we obtain

$$\cos 2\theta = C^2 - S^2 = 2C^2 - 1 = 1 - 2S^2,$$

using the identity $C^2 + S^2 = 1$. Hence

$$\cos 2\theta = \cos^2\theta - \sin^2\theta = 2\cos^2\theta - 1 = 1 - 2\sin^2\theta.$$

We also obtain similarly

$$\sin 2\theta = 2CS = 2\cos\theta\sin\theta.$$

Application 2 We can use the above formulae to obtain exact values for $\cos 45°$, $\sin 45°$ as follows. If we write $\theta = 45°$, $C = \cos 45°$, $S = \sin 45°$ then we have

$$0 = \cos 90° = 2C^2 - 1,$$

from which it follows that $2C^2 = 1$, and therefore $C^2 = 1/2$. Hence $C = \pm 1/\sqrt{2}$, which gives $\cos 45° = 1/\sqrt{2}$.

We also have $1 = \sin 90° = 2CS$, which gives $S = 1/2C = 1/\sqrt{2}$, and hence $\sin 45° = 1/\sqrt{2}$.

1.8 Euler's formulae for $\cos\theta$, $\sin\theta$ in terms of $e^{\pm i\theta}$

We obtained the formula $e^{i\theta} = \cos\theta + i\sin\theta$ in Section 1.6. From this formula we can derive two more formulae also attributed to Euler, viz.,

$$\cos\theta = \frac{e^{i\theta} + e^{-i\theta}}{2}, \quad \sin\theta = \frac{e^{i\theta} - e^{-i\theta}}{2i}.$$

Proof Observe that

$$\begin{aligned}e^{i\theta} &= \cos\theta + i\sin\theta,\\ e^{-i\theta} &= \cos\theta - i\sin\theta.\end{aligned}$$

Now eliminate $\sin\theta$, $\cos\theta$, respectively.

Application 3 We can use Euler's formulae to obtain formulae for $\cos^n \theta$, $\sin^n \theta$ in terms of $\cos k\theta$, $\sin k\theta$ $(0 \leq k \leq n)$. For example, we have

$$\cos^2 \theta = \left(\frac{e^{i\theta} + e^{-i\theta}}{2}\right)^2 = \frac{e^{2i\theta} + 2 + e^{-2i\theta}}{4} = \frac{1}{2}(1 + \cos 2\theta),$$

$$\sin^2 \theta = \left(\frac{e^{i\theta} - e^{-i\theta}}{2i}\right)^2 = \frac{e^{2i\theta} - 2 + e^{-2i\theta}}{-4} = \frac{1}{2}(1 - \cos 2\theta).$$

Application 4 Formulae of the above type are useful for integrating powers of $\cos\theta$, $\sin\theta$. For example,

$$\int \cos^2 \theta \, d\theta = \int \frac{1}{2}(1 + \cos 2\theta) \, d\theta = \frac{1}{2}\left(\theta + \frac{\sin 2\theta}{2}\right),$$

$$\int \sin^2 \theta \, d\theta = \int \frac{1}{2}(1 - \cos 2\theta) \, d\theta = \frac{1}{2}\left(\theta - \frac{\sin 2\theta}{2}\right).$$

1.9 *n*th roots

Suppose we have two complex numbers $z = re^{i\theta}$, $w = se^{i\phi}$. If we multiply them together we obtain

$$zw = rse^{i(\theta+\phi)},$$

which shows that $|zw| = |z|\,|w|$ as claimed in Section 1.2. Also that $\arg zw = \arg z + \arg w$. In particular, taking $z = w$ we have $z^2 = r^2 e^{2i\theta}$, and more generally $z^n = r^n e^{ni\theta}$. It follows that

$$z^{1/n} = r^{1/n} e^{i\theta/n}.$$

Observe that $r^{1/n}$ is the unique positive real nth root of r, whilst $e^{i\theta/n}$ has n possible values.

For example, if $z = -8$ then we have

$$z = 8e^{i\pi} = 8e^{3i\pi} = 8e^{5i\pi} = \cdots$$

$$z^{1/3} = 2e^{i\pi/3},\ 2e^{i\pi},\ 2e^{5i\pi/3}.$$

Even though $\arg(-8)$ has infinitely many values, there are only 3 distinct cube roots. We define the principal value of $(-8)^{1/3}$ to be that which corresponds to the principal value of $\arg(-8)$, namely π. So $(-8)^{1/3} = 2e^{i\pi/3}$ (PV).

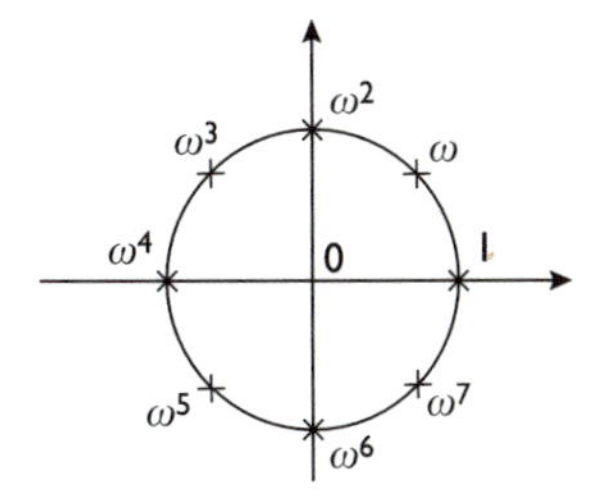

Figure 1.5

1.10 *n*th roots of unity

Just like any other non-zero complex number, 1 has n complex nth roots. We have

$$1 = e^0 = e^{2\pi i} = e^{4\pi i} = \cdots$$
$$1^{1/n} = e^0,\ e^{2\pi i/n},\ e^{4\pi i/n},\ \ldots$$

If we denote $\omega = e^{2\pi i/n}$, then the n nth roots of 1 are $1, \omega, \omega^2, \ldots, \omega^{n-1}$ (see Figure 1.5 for the case $n = 8$). We call ω the *primitive* nth root of 1.

N.B. $1^{1/n} = 1$ (PV) of course.

Lemma $1 + \omega + \omega^2 + \cdots + \omega^{n-1} = 0$.

Proof $1 + \omega + \omega^2 + \cdots + \omega^{n-1} = \dfrac{1 - \omega^n}{1 - \omega} = 0.$

1.11 Inequalities

The fundamental inequality is the so called triangle, or parallelogram inequality and is as follows.

Inequality 1 $|z + w| \le |z| + |w|$. This inequality expresses the fact that the diagonal of a parallelogram has length less than or equal to the sum of the lengths of two adjacent sides (see Figure 1.6). Equivalently, that the length of one side of a triangle is less than or equal to the sum of the lengths of the other two sides. (Consider the triangle with vertices $0, z, z + w$.)

Inequality 2 $|z - w| \le |z| + |w|$. This inequality follows from Inequality 1 by putting $-w$ for w.

N.B. Note the plus sign on the right-hand side.

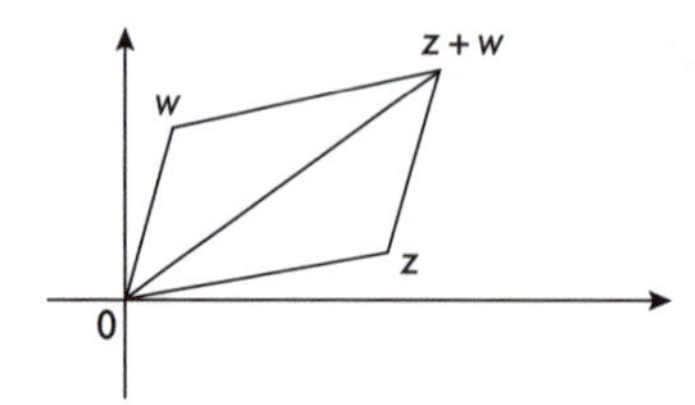

Figure 1.6

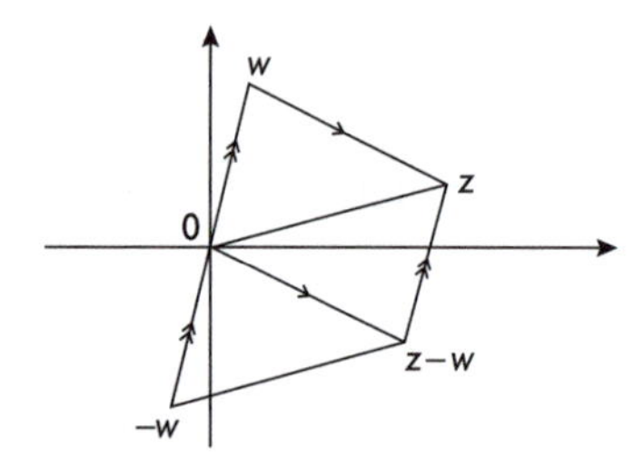

Figure 1.7

Note also that $|z - w|$ has a geometric significance as the distance between z and w on the Argand diagram (see Figure 1.7).

Inequality 3 $|z - w| \geq |z| - |w|$. This inequality follows from Inequality 1 by observing that

$$|z| = |(z - w) + w| \leq |z - w| + |w|.$$

Inequality 4 $|z - w| \geq |w| - |z|$. Observe that

$$|z - w| = |w - z| \geq |w| - |z|.$$

Inequality 5 $|z + w| \geq |z| - |w|$. Put $-w$ for w in Inequality 3.

N.B. Note the minus sign on the right-hand side.

Worked Example Prove that for all $|z| = 2$

$$\frac{1}{4} \leq \left|\frac{z^2 + 1}{z^2 + 8}\right| \leq \frac{5}{4}.$$

Solution To prove the right-hand inequality we observe first that

$$|z^2 + 1| \leq |z^2| + |1| = 4 + 1 = 5,$$

and second that

$$|z^2 + 8| \geq 8 - |z^2| = 8 - 4 = 4.$$

The left-hand inequality is proved similarly.

1.12 Extension to 3 terms (or more)

We give the inequalities for 3 terms. The generalization to more terms is left to the reader.

Inequality 6 $\quad |A + B + C| \leq |A| + |B| + |C|$.

Proof Observe that

$$|A + B + C| \leq |A + B| + |C| \leq |A| + |B| + |C|$$

by repeated application of Inequality 1.

Inequality 7 $\quad |A + B + C| \geq |A| - |B| - |C|$.

Proof Observe that

$$\begin{aligned} |A| &= |(A + B + C) - B - C| \\ &\leq |A + B + C| + |-B| + |-C| \\ &= |A + B + C| + |B| + |C|. \end{aligned}$$

Notes

We never defined $e^{i\theta}$, or proved that the laws of indices hold for complex exponents. A rigorous treatment of this material would *define* e^z, $\cos z$, $\sin z$ by their Maclaurin series

$$\begin{aligned} e^z &= 1 + z + \frac{z^2}{2!} + \frac{z^3}{3!} + \cdots \\ \cos z &= 1 - \frac{z^2}{2!} + \frac{z^4}{4!} - \frac{z^6}{6!} + \cdots \\ \sin z &= z - \frac{z^3}{3!} + \frac{z^5}{5!} - \frac{z^7}{7!} + \cdots \end{aligned}$$

and prove their properties by manipulation of these series.

Purists might prefer to prove $|zw| = |z|\,|w|$ and $|z+w| \le |z|+|w|$ by algebraic methods. For example, if $z = x + iy$, $w = u + iv$, then we have

$$zw = (x + iy)(u + iv) = (xu - yv) + i(xv + yu).$$

Therefore

$$\begin{aligned}
|zw|^2 &= (xu - yv)^2 + (xv + yu)^2 \\
&= (x^2u^2 - 2xuyv + y^2v^2) + (x^2v^2 + 2xvyu + y^2u^2) \\
&= x^2u^2 + y^2v^2 + x^2v^2 + y^2u^2 \\
&= (x^2 + y^2)(u^2 + v^2) \\
&= |z|^2|w|^2.
\end{aligned}$$

We also have $z + w = (x + u) + i(y + v)$. Therefore,

$$\begin{aligned}
|z + w|^2 &= (x + u)^2 + (y + v)^2 = (x^2 + 2xu + u^2) + (y^2 + 2yv + v^2), \\
(|z| + |w|)^2 &= |z|^2 + 2|zw| + |w|^2 \\
&= x^2 + y^2 + u^2 + v^2 + 2\sqrt{(x^2 + y^2)(u^2 + v^2)}.
\end{aligned}$$

From which it follows that

$$(|z| + |w|)^2 - |z + w|^2 = 2\sqrt{(x^2 + y^2)(u^2 + v^2)} - 2xu - 2yv,$$

which is ≥ 0 if

$$(xu + yv)^2 \le (x^2 + y^2)(u^2 + v^2).$$

However,

$$\begin{aligned}
(x^2 + y^2)(u^2 + v^2) - (xu + yv)^2 &= (x^2u^2 + x^2v^2 + y^2u^2 + y^2v^2) \\
&\quad - (x^2u^2 + 2xuyv + y^2v^2) \\
&= x^2v^2 + y^2u^2 - 2xuyv \\
&= (xv - yu)^2 \\
&\ge 0
\end{aligned}$$

as required.

Examples

1. Express the following complex numbers in the form $x + iy$.

   (i) $(1 + 3i) + (5 + 7i)$, (ii) $(1 + 3i) - (5 + 7i)$, (iii) $(1 + 3i)(5 + 7i)$,

   (iv) $\dfrac{1 + 3i}{5 + 7i}$, (v) $\sqrt{3 + 4i}$, (vi) $\log(1 + i)$.

   Hint For (vi) use the polar form.

2. Find $\sqrt{1 + i}$. Hence show $\tan \pi/8 = \sqrt{2} - 1$.

3. Expand $(\cos\theta + i \sin\theta)^3$ to obtain formulae for $\cos 3\theta$, $\sin 3\theta$ in terms of $\cos\theta$, $\sin\theta$. Use these formulae to show

$$\cos 3\theta = 4\cos^3\theta - 3\cos\theta,$$

$$\sin 3\theta = 3\sin\theta - 4\sin^3\theta.$$

4. Use Question 3 to show that $\cos 30° = \sqrt{3}/2$, $\sin 30° = 1/2$.

5. Expand $(e^{i\theta} \pm e^{-i\theta})^3$ to show

$$\cos^3\theta = \tfrac{1}{4}(3\cos\theta + \cos 3\theta),$$

$$\sin^3\theta = \tfrac{1}{4}(3\sin\theta - \sin 3\theta).$$

6. Use Question 5 to evaluate $\displaystyle\int_0^{\pi/2} \cos^3\theta \, d\theta$, $\displaystyle\int_0^{\pi/2} \sin^3\theta \, d\theta$.

7. Evaluate the integral $\displaystyle\int_0^{\pi} e^{2x} \cos 4x \, dx$ by taking the real part of

$$\int_0^{\pi} e^{2x} e^{4ix} \, dx = \int_0^{\pi} e^{(2+4i)x} \, dx.$$

   Now do it by integrating by parts twice, and compare the efficiency of the two methods.

Chapter 2

Complex functions

2.1 Polynomials

Having constructed the complex number system the next task is to consider how the standard functions we do real calculus with extend to complex variables. Polynomials cause no problems since they only require addition, multiplication and subtraction for their definition. For example, $p(z) = 3z + 4$, $q(z) = 4z^2 - 5z + 6$, etc. The numbers occurring are called *coefficients*. The *degree* of the polynomial is the highest power of z occurring with a non-zero coefficient.

2.2 Rational functions

These are functions of the form $r(z) = p(z)/q(z)$ where $p(z), q(z)$ are polynomials. They can be defined for all z except where the denominator vanishes. Such points are called *singularities*. Every rational function has at least one singularity because of the fundamental theorem of algebra. For example,

$$r(z) = \frac{z+1}{z+2}$$

has a singularity at $z = -2$, whilst

$$s(z) = \frac{z^2+1}{z^2+4}$$

has two singularities at $z = \pm 2i$.

2.3 Graphs

Every real function $y = f(x)$ of a real variable x has a graph in two dimensional space. For example, Figure 2.1 shows the graph of $y = x^2$.

For a complex function $w = f(z)$ of a complex variable z this option is not available because the graph is a two-dimensional surface in a four-dimensional space. What we have to do instead is to draw two diagrams which we call a z-plane and

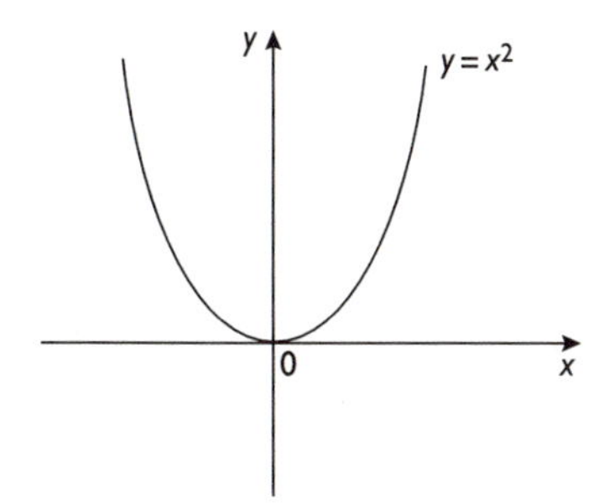

Figure 2.1

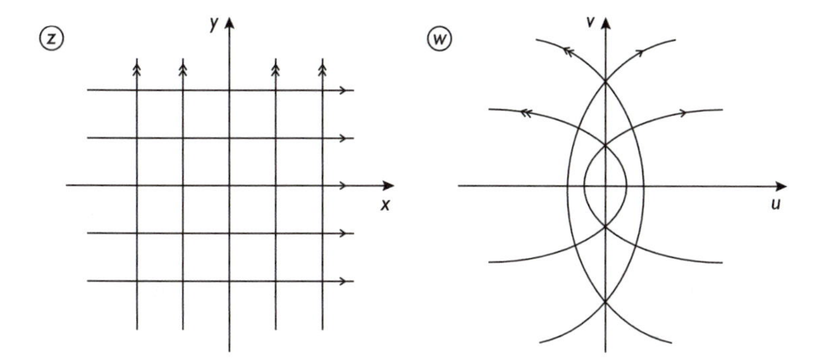

Figure 2.2

a w-plane, and then indicate how geometrical figures in the z-plane are transformed to geometrical figures in the w-plane under the action of the function $w = f(z)$.

For example, for the complex function $w = z^2$ we find that the grid lines $x =$ constant, $y =$ constant in the z-plane transform to confocal parabolas in the w-plane (Figure 2.2).

To see this observe that if $z = x + iy$, $w = u + iv$, then

$$u + iv = (x + iy)^2 = x^2 - y^2 + 2ixy,$$

therefore

$$u = x^2 - y^2, \quad v = 2xy,$$

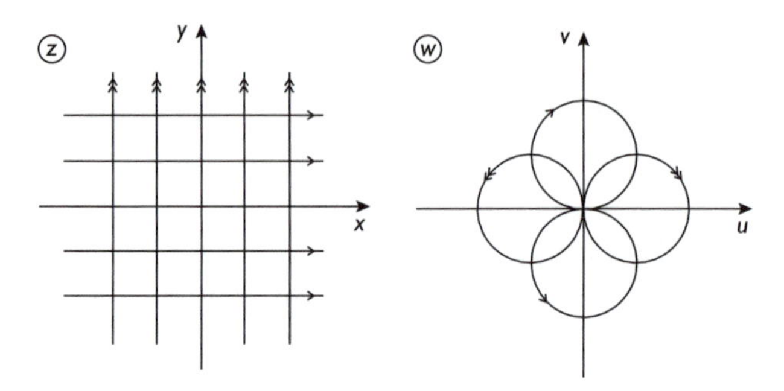

Figure 2.3

which gives, on eliminating y,

$$u = x^2 - \frac{v^2}{4x^2},$$

$$4x^4 - 4x^2u - v^2 = 0,$$

$$u^2 + v^2 = u^2 - 4x^2u + 4x^4 = (u - 2x^2)^2,$$

$$|w| = |\text{Re}\, w - 2x^2|,$$

which is the equation of a parabola with focus $w = 0$, directrix $\text{Re}\, w = 2x^2$. This parabola is the image of the line $x =$ constant.

Similarly, eliminating x, we get

$$u^2 + v^2 = (u + 2y^2)^2,$$

$$|w| = |\text{Re}\, w + 2y^2|,$$

which is a parabola, again with focus $w = 0$, but now with directrix $\text{Re}\, w = -2y^2$. This is the image of the grid line $y =$ constant.

Another example which readers might like to work out for themselves is $w = 1/z$ which transforms the grid lines $x =$ constant, $y =$ constant in the z-plane to circles through the origin with centres on the real and imaginary axes in the w-plane (see Figure 2.3).

2.4 The exponential function

For real variables the function $y = e^x$ has the graph illustrated in Figure 2.4.

For complex variables we have

$$w = e^z = e^{x+iy} = e^x e^{iy}$$

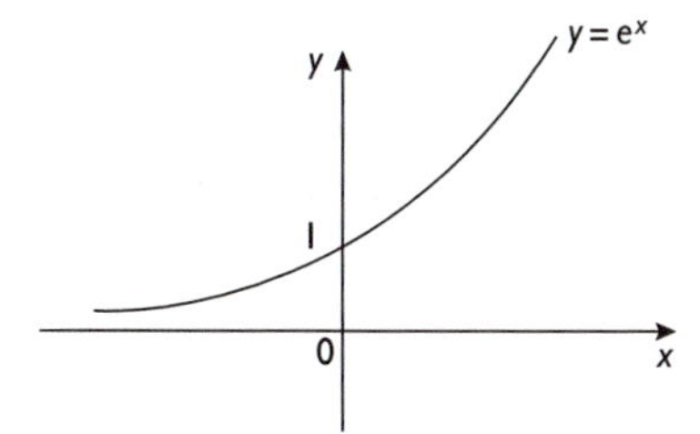

Figure 2.4

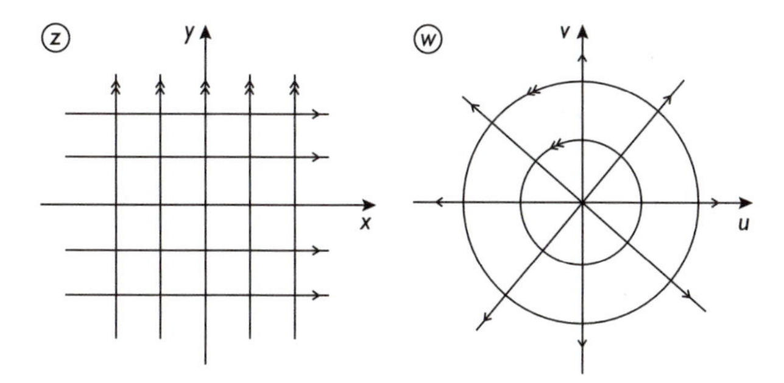

Figure 2.5

showing that if we use the polar form $w = se^{i\phi}$ we get $s = e^x, \phi = y$. In other words

$$|e^z| = e^{\operatorname{Re} z}, \quad \arg e^z = \operatorname{Im} z.$$

This will of course not be the principal value of $\arg e^z$ unless $-\pi < \operatorname{Im} z \leq \pi$.

The complex graph of $w = e^z$ is as in Figure 2.5. The grid lines $x =$ constant go to circles centre the origin. The grid lines $y =$ constant go to half lines emanating from the origin.

2.5 Trigonometric and hyperbolic functions

For real variables the trigonometric functions and the hyperbolic functions are very different animals. For example, the graphs for $\sin x$, $\cos x$ are periodic and bounded (see Figure 2.6). Whereas the graphs for $\sinh x$, $\cosh x$ are neither periodic nor bounded (see Figure 2.7).

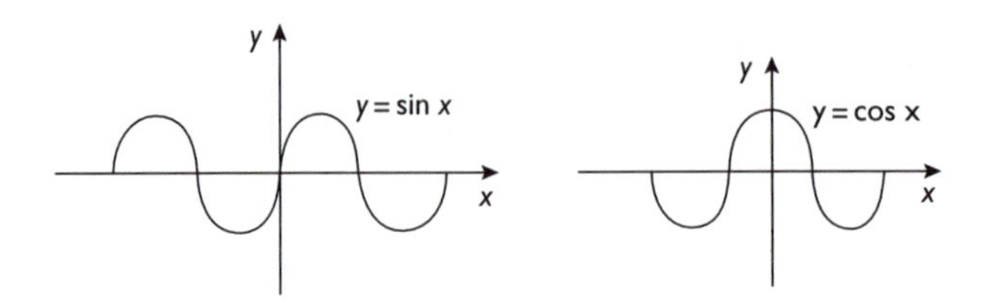

Figure 2.6

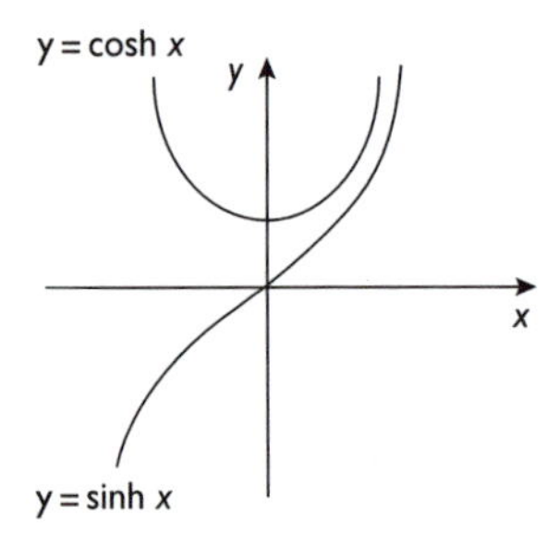

Figure 2.7

For complex variables however it turns out that trigonometric functions and hyperbolic functions are intimately related. The following formulae are fundamental for all that follows.

2.6 Fundamental formulae

For all real x we have

$$\begin{aligned}
\sin(ix) &= i \sinh x, \\
\sinh(ix) &= i \sin x, \\
\cos(ix) &= \cosh x, \\
\cosh(ix) &= \cos x.
\end{aligned}$$

These formulae can be proved in several ways. For example, by definition

$$\begin{aligned}
\sinh(ix) &= \frac{e^{ix} - e^{-ix}}{2} = i \sin x, \\
\cosh(ix) &= \frac{e^{ix} + e^{-ix}}{2} = \cos x
\end{aligned}$$

from Euler's formulae for $\sin x$, $\cos x$ (see Section 1.8).

Or, from the Maclaurin series we have

$$\begin{aligned}\sin(ix) &= ix - \frac{(ix)^3}{3!} + \frac{(ix)^5}{5!} - \cdots \\ &= ix + i\frac{x^3}{3!} + i\frac{x^5}{5!} + \cdots \\ &= i \sinh x, \\ \cos(ix) &= 1 - \frac{(ix)^2}{2!} + \frac{(ix)^4}{4!} - \cdots \\ &= 1 + \frac{x^2}{2!} + \frac{x^4}{4!} + \cdots \\ &= \cosh x.\end{aligned}$$

2.7 Application I

We can use the Fundamental formulae of 2.6 to obtain the real and imaginary parts of $\sin z$, and hence draw the graph of $w = \sin z$. If we write $z = x+iy$, $w = u+iv$, then we have

$$\begin{aligned}\sin(x + iy) &= \sin x \cos(iy) + \cos x \sin(iy) \\ &= \sin x \cosh y + i \cos x \sinh y,\end{aligned}$$

which gives

$$u = \sin x \cosh y, \quad v = \cos x \sinh y.$$

Eliminating x we get

$$\frac{u^2}{\cosh^2 y} + \frac{v^2}{\sinh^2 y} = 1$$

which is the equation of an ellipse with foci at ± 1. Eliminating y we get

$$\frac{u^2}{\sin^2 x} - \frac{v^2}{\cos^2 x} = 1$$

which is the equation of a hyperbola with foci at ± 1.

It follows that $w = \sin z$ transforms the grid lines $x =$ constant, $y =$ constant in the z-plane to confocal ellipses and hyperbolae in the w-plane (see Figure 2.8).

The graph of $w = \cos z$ is similar. For $\sinh z$, $\cosh z$ we also get confocal ellipses and hyperbolae, but with foci at $\pm i$ instead of at ± 1.

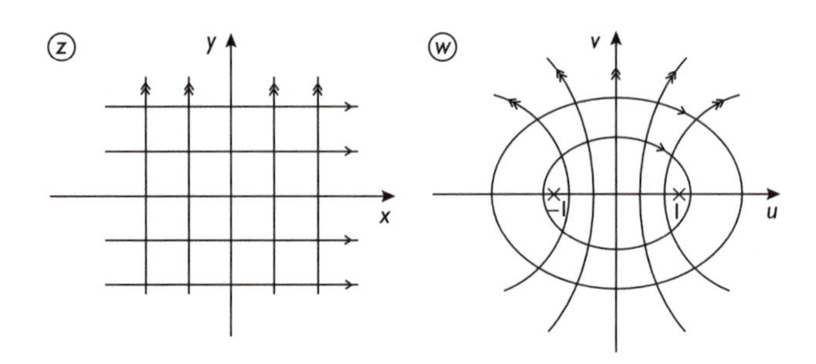

Figure 2.8

2.8 Application 2

The inequality $|\sin x| \leq 1$ for real x fails for complex variables. If we write $z = x + iy$, then we have

$$
\begin{aligned}
|\sin z|^2 = |\sin(x+iy)|^2 &= \sin^2 x \cosh^2 y + \cos^2 x \sinh^2 y \\
&= \sin^2 x(1+\sinh^2 y) + (1-\sin^2 x)\sinh^2 y = \sin^2 x + \sinh^2 y.
\end{aligned}
$$

So if, for example, $z = \pi/2 + i\epsilon$, where $\epsilon > 0$, then $|\sin z|^2 = 1 + \sinh^2 \epsilon > 1$.

2.9 Application 3

The only zeros of $\sin z$ for complex z are the real zeros at $z = n\pi$ for integral n. This is because if $z = x + iy$ and $\sin z = 0$ then

$$0 = |\sin z|^2 = \sin^2 x + \sinh^2 y.$$

Therefore $\sin x = \sinh y = 0$, which gives $x = n\pi$, $y = 0$ and hence $z = n\pi$.

Similarly, we leave it as an exercise for the reader to show that the only zeros of $\cos z$ for complex z are at $z = n\pi + \pi/2$ for integral n.

2.10 Identities for hyperbolic functions

The fundamental formulae (see Section 2.6) can be used to obtain identities for hyperbolic functions from analogous identities for trigonometric functions. For example, the trigonometric identity $\sin^2 x + \cos^2 x = 1$ gives, on substituting ix for x,

$$1 = \sin^2(ix) + \cos^2(ix) = (i\sinh x)^2 + (\cosh x)^2 = \cosh^2 x - \sinh^2 x.$$

2.11 The other trigonometric functions

We define $\tan z$, $\cot z$, $\sec z$, $\operatorname{cosec} z$ in terms of $\sin z$, $\cos z$ as follows.

$$\tan z = \frac{\sin z}{\cos z}, \quad \cot z = \frac{\cos z}{\sin z}, \quad \sec z = \frac{1}{\cos z}, \quad \operatorname{cosec} z = \frac{1}{\sin z}.$$

Similarly for the other hyperbolic functions.

These functions all have singularities. For example, $\tan z$ has singularities at the zeros of $\cos z$, that is, $z = n\pi + \pi/2$. The corresponding hyperbolic function $\tanh z = \sinh z/\cosh z$ has singularities at the zeros of $\cosh z$, that is, $z = i(n\pi + \pi/2)$.

2.12 The logarithmic function

The graph of $y = \log x$ for real x is as in Figure 2.9. Observe that $\log x$ is only defined for $x > 0$. This is because the real exponential function only takes positive values (see Figure 2.4).

To define $\log z$ for complex z we use the polar form $z = re^{i\theta}$. We get

$$\log z = \log(re^{i\theta}) = \log r + \log(e^{i\theta}) = \log r + i\theta = \log|z| + i\arg z.$$

Since $\arg z$ is many valued it follows that $\log z$ is also many valued. We define the *principal value* of $\log z$ to be the one obtained by taking the principal value of $\arg z$. For example, we have $1 + i = \sqrt{2}e^{i\pi/4}$ (PV) therefore

$$\log(1+i) = \frac{1}{2}\log 2 + i\frac{\pi}{4} \text{ (PV)}.$$

Observe that $\log z$ has a singularity at $z = 0$ since we cannot define $\log r$ for $r = 0$.

To get the complex graph for $w = \log z$ it is best to consider the action of $\log z$ on the circles $|z| =$ constant and the half lines $\arg z =$ constant in the z-plane. These transform to the grid lines $\operatorname{Re} w =$ constant, $\operatorname{Im} w =$ constant in the w-plane (see Figure 2.10).

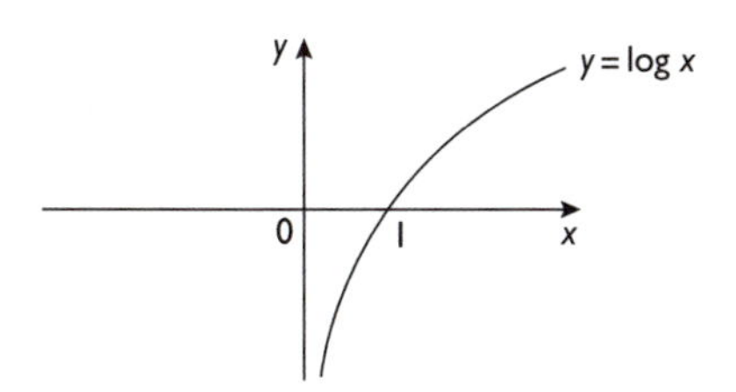

Figure 2.9

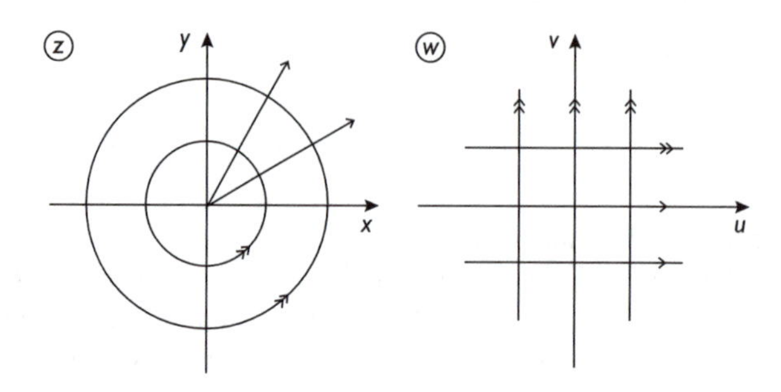

Figure 2.10

Notes

We have not actually defined e^z, $\sin z$, $\cos z$, $\log z$ for complex z. We have merely assumed that these functions can be defined, and that they continue to have the properties they possess in the real domain. For example, laws of indices, laws of logarithms, trigonometric identities. A rigorous treatment would define e^z, $\sin z$, $\cos z$, $\log z$ from their Maclaurin series, and derive their properties from these series. The function $\log z$ would be defined as the inverse function of e^z.

The functions we have drawn complex graphs of are all conformal mappings in the sense that curves which intersect at an angle θ in the z-plane transform to curves which intersect at the same angle θ in the w-plane. Observe that in every case the grid lines $x =$ constant, $y =$ constant in the z-plane transform to curves which intersect orthogonally in the w-plane. This conformal property is crucial in applications to fluid dynamics.

Examples

1. Prove that for all $|z| = 2$

$$2 \leq |z - 4| \leq 6.$$

2. Prove that for all $|z| = 3$

$$\frac{8}{11} \leq \left|\frac{z^2 + 1}{z^2 + 2}\right| \leq \frac{10}{7}.$$

3. Prove that for all $|z| = 4$

$$\frac{3}{5} \leq \left|\frac{z + i}{z - i}\right| \leq \frac{5}{3}.$$

4. Prove that for all $|z| = R > 2$

$$\left|\frac{1}{z^2 + z + 1}\right| \leq \frac{1}{R^2 - R - 1}.$$

5. Prove that $|e^z| = e^{\operatorname{Re} z}$.
6. Find where $|e^z|$ is maximum for $|z| \leq 2$ (draw a diagram).
7. Prove that for $z = x + iy$

$$|\sin(x + iy)|^2 = \sin^2 x + \sinh^2 y,$$

$$|\cos(x + iy)|^2 = \cos^2 x + \sinh^2 y.$$

8. Find where $|\sin z|$ is maximum for $|z| \leq 1$ (draw a diagram).
9. Prove that all points z satisfying

$$\left|\frac{z + 1}{z + 4}\right| = 2$$

lie on a circle. Find its centre and radius.

Chapter 3

Derivatives

3.1 Differentiability and continuity

For a real function $f(x)$ of a real variable x the *derivative* $f'(x)$ is defined as the limit

$$f'(x) = \lim_{h\to 0} \frac{f(x+h) - f(x)}{h}.$$

Observe that (see Figure 3.1)

$$\frac{f(x+h) - f(x)}{h}$$

is the gradient of the line PQ which converges to the tangent at P as $Q \to P$. So $f'(x)$ is the gradient of the tangent at P.

For example, if $f(x) = x^2$ then we have

$$\begin{aligned}\frac{f(x+h) - f(x)}{h} &= \frac{(x+h)^2 - x^2}{h} = \frac{x^2 + 2xh + h^2 - x^2}{h} \\ &= \frac{2xh + h^2}{h} = 2x + h,\end{aligned}$$

which $\to 2x$ as $h \to 0$. Therefore $f'(x) = 2x$.

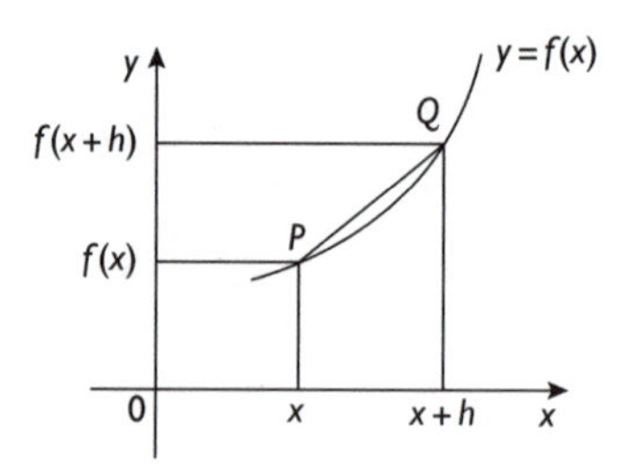

Figure 3.1

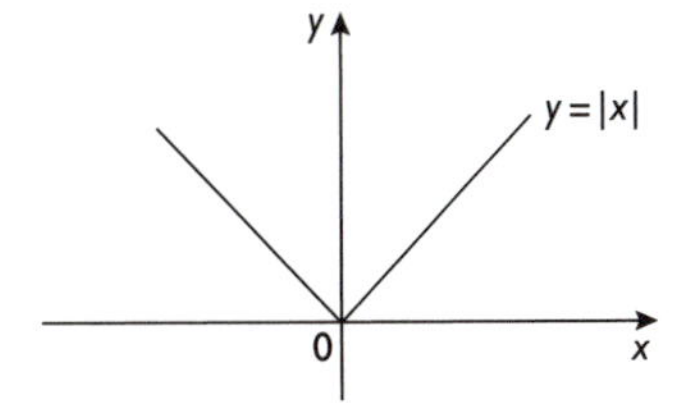

Figure 3.2

Similarly one can in principle go through all the elementary functions of calculus and show they have the derivatives they are supposed to have.

We can also prove all the elementary combination rules for differentiating sums, products, quotients and composites.

We cannot assume that the derivative $f'(x)$ always exists. For example, if $f(x) = |x|$ then

$$\begin{aligned} \frac{f(h) - f(0)}{h} &= \frac{h}{h} = 1 \quad (h > 0) \\ &= \frac{|h|}{h} = -1 \quad (h < 0), \end{aligned}$$

so has no limit as $h \to 0$.

Observe that the graph of $f(x) = |x|$ has no well defined tangent at $x = 0$ (see Figure 3.2).

We therefore define $f(x)$ to be *differentiable* at x if

$$\lim_{h \to 0} \frac{f(x+h) - f(x)}{h}$$

exists. According to this definition $f(x) = |x|$ is not differentiable at $x = 0$.

Another case where differentiability fails is at a discontinuity of $f(x)$. A *continuous* function $f(x)$ is one whose graph has no breaks. We make this idea precise by defining $f(x)$ to be *continuous* at x if

$$\lim_{h \to 0} f(x+h) = f(x).$$

For example, $f(x) = 1/x$ is not continuous at $x = 0$ (Figure 3.3). In this connection we have the following theorem.

Theorem 1 Differentiability implies continuity.

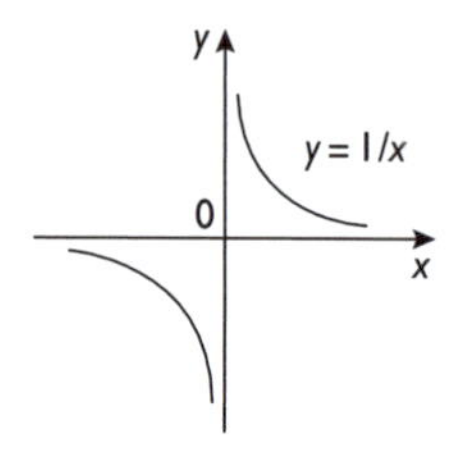

Figure 3.3

Proof Suppose $f(x)$ is differentiable at x, then we have

$$f(x+h) - f(x) = h\frac{f(x+h) - f(x)}{h} \to 0 \times f'(x) = 0$$

as $h \to 0$. Therefore $f(x+h) \to f(x)$ as $h \to 0$.

Corollary $f(x) = 1/x$ is not differentiable at $x = 0$.

Observe that the converse of Theorem 1 is false. A counterexample is $f(x) = |x|$ which is continuous but not differentiable at $x = 0$.

For a complex function of a complex variable z, we define differentiability and continuity of $f(z)$ exactly as we have done for real functions of a real variable. The familiar functions all have their familiar derivatives, and the familiar combination rules are all valid. There is also a further constraint in the form of the Cauchy–Riemann equations to which we devote the next section.

3.2 Cauchy–Riemann equations

Suppose we have a complex valued function $w = f(z)$ of the complex variable z, and suppose we write $w = u + iv$, $z = x + iy$, then we can express u, v as functions of x, y and consider their partial derivatives $\partial u/\partial x$, $\partial u/\partial y$, $\partial v/\partial x$, $\partial v/\partial y$. For example, if $w = z^2$, then

$$u + iv = w = z^2 = (x + iy)^2 = x^2 + 2ixy - y^2,$$

which gives in this case

$$u = x^2 - y^2, \quad v = 2xy.$$

In general we can use the chain rule to obtain

$$\frac{\partial w}{\partial x} = \frac{dw}{dz}\frac{\partial z}{\partial x} = \frac{\partial u}{\partial x} + i\frac{\partial v}{\partial x},$$

$$\frac{\partial w}{\partial y} = \frac{dw}{dz}\frac{\partial z}{\partial y} = \frac{\partial u}{\partial y} + i\frac{\partial v}{\partial y},$$

which, on observing that $\partial z/\partial x = 1$, $\partial z/\partial y = i$, gives

$$\frac{dw}{dz} = \frac{\partial u}{\partial x} + i\frac{\partial v}{\partial x} = -i\frac{\partial u}{\partial y} + \frac{\partial v}{\partial y}.$$

Therefore on equating real and imaginary parts we have

$$\frac{\partial u}{\partial x} = \frac{\partial v}{\partial y}, \quad \frac{\partial v}{\partial x} = -\frac{\partial u}{\partial y}.$$

These are the *Cauchy–Riemann* equations published independently by Cauchy (1818) and Riemann (1851).

We call the formula

$$\frac{dw}{dz} = \frac{\partial u}{\partial x} + i\frac{\partial v}{\partial x}$$

the *Cauchy–Riemann* formula for the derivative.

In the case $w = z^2$, we get

$$\frac{\partial u}{\partial x} = \frac{\partial v}{\partial y} = 2x, \quad \frac{\partial v}{\partial x} = -\frac{\partial u}{\partial y} = 2y.$$

Also the Cauchy–Riemann formula gives

$$\frac{dw}{dz} = \frac{\partial u}{\partial x} + i\frac{\partial v}{\partial x} = 2x + 2iy = 2z$$

as expected.

3.3 Failure of the Cauchy–Riemann equations

Consider the function $w = \bar{z} = x - iy$. If $w = u + iv$, then we have $u = x$, $v = -y$. Therefore

$$\frac{\partial u}{\partial x} = 1, \quad \frac{\partial v}{\partial y} = -1, \quad \frac{\partial v}{\partial x} = -\frac{\partial u}{\partial y} = 0,$$

which means that the first Cauchy–Riemann equation is not satisfied for any x, y. We are forced to the conclusion that the function $f(z) = \bar{z}$ cannot be differentiable for any z.

In this connection we have the following theorem.

Theorem 2 For $u + iv = f(x + iy)$ with continuous partial derivatives $\partial u/\partial x$, $\partial u/\partial y$, $\partial v/\partial x$, $\partial v/\partial y$ the function $f(z)$ is differentiable at z if and only if the Cauchy–Riemann equations

$$\frac{\partial u}{\partial x} = \frac{\partial v}{\partial y}, \quad \frac{\partial v}{\partial x} = -\frac{\partial u}{\partial y},$$

are satisfied.

Proof We proved necessity above. For sufficiency we refer the reader to rigorous books on complex analysis.

3.4 Geometric significance of the complex derivative

For a real function $f(x)$ of a real variable x, the equation of the tangent to the graph $y = f(x)$ at $x = a$ is

$$y = f(a) + (x - a)f'(a).$$

For a complex function $f(z)$ of a complex variable z, the equation of the tangent plane (in 4 dimensions) to the graph $w = f(z)$ at $z = a$ is

$$w = f(a) + (z - a)f'(a) = Az + B,$$

where $A = f'(a)$, $B = f(a) - af'(a)$.

The geometric effect of the linear function $w = Az + B$ is a rotation, a scaling, and a translation. The rotation is through the angle $\arg A$, the scaling is by the factor $|A|$. The translation is through a distance $|B|$ in the direction $\arg B$.

What this tells us about the transformation $w = f(z)$ is that near $z = a$ the effect is approximately a rotation through $\arg f'(a)$, and a scaling by $|f'(a)|$. For example, if $a = i\pi/2$ and $f(z) = e^z$, then we have $f(a) = e^{i\pi/2} = i$. Also $f'(a) = e^a = e^{i\pi/2} = i$. So the effect near $z = a$ is a rotation through 90° anticlockwise (see Figure 3.4). If $b = i\pi/2 + 1$, then we have $f(b) = f'(b) = ei$, so the effect locally is now a scaling by e, and again a rotation through 90° anticlockwise (see Figure 3.4).

The fact that $w = f(z)$ acts locally like a rotation through $\arg f'(z)$ explains why curves which intersect at a certain angle in the z-plane are transformed under the action of $w = f(z)$ to curves which intersect at the same angle in the w-plane. This is the characteristic property of a *conformal mapping* which is important for the applications to fluid mechanics.

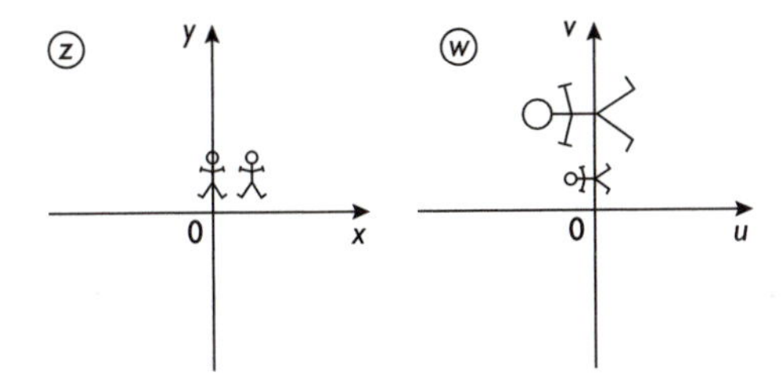

Figure 3.4

3.5 Maclaurin expansions

It has always been important to be able to approximate functions by polynomials. This is because polynomials are the only functions whose values can be calculated arithmetically. For a calculator to calculate e^x for given x it has to evaluate the series

$$e^x = 1 + x + \frac{x^2}{2!} + \frac{x^3}{3!} + \cdots$$

to as many terms as are needed to achieve the required degree of accuracy. To calculate the value of π it is necessary to use the series

$$\tan^{-1} x = x - \frac{x^3}{3} + \frac{x^5}{5} - \cdots$$

with $x = 1$. In practice, both of these calculations are done by more sophisticated methods, but they still have to make use of polynomial expansions in one form or another.

Maclaurin (1742) gave the general form for expanding a function $f(x)$ in powers of x. The expansion is

$$f(x) = \sum_{n=0}^{\infty} a_n x^n,$$

where the nth coefficient a_n is given by the formula

$$a_n = \frac{f^{(n)}(0)}{n!},$$

and where $f^{(n)}(0)$ denotes the nth derivative $f^{(n)}(x)$ of $f(x)$ evaluated at $x = 0$. We call this expansion the *Maclaurin* expansion of $f(x)$, and we call the coefficient a_n the nth *Maclaurin* coefficient of $f(x)$.

For example, if $f(x) = e^x$, then we have $f'(x) = f''(x) = \cdots = e^x$, and therefore $f'(0) = f''(0) = \cdots = 1$. So the Maclaurin expansion of e^x is

$$e^x = 1 + x + \frac{x^2}{2!} + \frac{x^3}{3!} + \cdots$$

as already observed above.

To see where the Maclaurin formula for the nth coefficient comes from, observe that if

$$f(x) = a_0 + a_1 x + a_2 x^2 + a_3 x^3 + \cdots$$

then putting $x = 0$ gives $f(0) = a_0$.

Differentiating term by term we get

$$f'(x) = a_1 + 2a_2 x + 3a_3 x^2 + \cdots,$$

which on substituting $x = 0$ gives $f'(0) = a_1$.

Differentiating again we get

$$f''(x) = 2a_2 + 6a_3 x^2 + \cdots,$$

which on substituting $x = 0$ gives $f''(0) = 2a_2$, and hence $a_2 = f''(0)/2!$

Similarly, differentiating n times and putting $x = 0$ we get $f(n)(0) = n!a_n$, and hence $a_n = f(n)(0)/n!$ as required.

Maclaurin was concerned with real variables only, but his expansion remains valid for complex variables also. We list below some examples of Maclaurin expansions in the complex context.

$$e^z = 1 + z + \frac{z^2}{2!} + \frac{z^3}{3!} + \cdots$$

$$\sin z = z - \frac{z^3}{3!} + \frac{z^5}{5!} - \cdots$$

$$\cos z = 1 - \frac{z^2}{2!} + \frac{z^4}{4!} - \cdots$$

$$\sinh z = z + \frac{z^3}{3!} + \frac{z^5}{5!} + \cdots$$

$$\cosh z = 1 + \frac{z^2}{2!} + \frac{z^4}{4!} + \cdots$$

$$(1+z)^\alpha = 1 + \alpha z + \frac{\alpha(\alpha-1)}{2!} z^2 + \cdots \quad (|z| < 1)$$

$$\frac{1}{1-z} = 1 + z + z^2 + \cdots \quad (|z| < 1)$$

$$\log(1+z) = z - \frac{z^2}{2} + \frac{z^3}{3} - \cdots \quad (|z| < 1).$$

The first five expansions are valid for all z, whilst the last three are only valid for $|z| < 1$. The expansion for $(1+z)^\alpha$ is of course the binomial theorem, which gives a terminating series in the case α a positive integer. The particular case $\alpha = -1$ gives the geometric series

$$\frac{1}{1+z} = 1 - z + z^2 - \cdots,$$

which on integrating term by term gives the series for $\log(1+z)$ (PV).

3.6 Calculating Maclaurin expansions

We can either use the Maclaurin formula $a_n = f^{(n)}(0)/n!$ or we can combine the standard expansions listed in Section 3.5.

For example, suppose $f(z) = \tan z$. Then writing $T = \tan z$, $S = \sec z$ and observing that $dT/dz = S^2$, $dS/dz = ST$ we have the following.

$$f(z) = T = 0 \text{ at } z = 0, \text{ therefore } a_0 = 0.$$

$$f'(z) = S^2 = 1 \text{ at } z = 0, \text{ therefore } a_1 = 1.$$

$$f''(z) = 2S^2T = 0 \text{ at } z = 0, \text{ therefore } a_2 = 0.$$

$$f'''(z) = 4S^2T^2 + 2S^4 = 2 \text{ at } z = 0, \text{ therefore } a_3 = 2/3! = 1/3.$$

Hence we obtain

$$\tan z = z + \frac{z^3}{3} + \cdots$$

Alternatively, we can write

$$\begin{aligned}
\tan z = \frac{\sin z}{\cos z} &= \frac{z - z^3/3! + z^5/5! - \cdots}{1 - z^2/2! + z^4/4! - \cdots} \\
&= \left(z - \frac{z^3}{6} + \frac{z^5}{120} - \cdots\right) \\
&\quad \times \left(1 + \left(\frac{z^2}{2} - \frac{z^2}{24} + \cdots\right) + \left(\frac{z^2}{2} - \frac{z^2}{24} + \cdots\right)^2 - \cdots\right) \\
&= \left(z - \frac{z^3}{6} + \frac{z^5}{120} - \cdots\right)\left(1 + \frac{z^2}{2} + \frac{5}{24}z^4 + \cdots\right) \\
&= z + \frac{z^3}{3} + \frac{2}{15}z^5 + \cdots.
\end{aligned}$$

3.7 Taylor expansions

The Maclaurin expansion is a particular case of a more general expansion due to Taylor (1715) which represents $f(z)$ as a series in powers of $z-c$ for any fixed c as

$$f(z) = \sum_{n=0}^{\infty} a_n (z-c)^n,$$

where the nth coefficient a_n is given by the formula

$$a_n = \frac{f^{(n)}(c)}{n!}.$$

We call this expansion the *Taylor* expansion of $f(z)$ at $z = c$, and we call the coefficient a_n the nth *Taylor* coefficient of $f(z)$ at $z = c$.

For example, suppose $f(z) = 1/z$ and $c = 1$. We can calculate a_n as follows:

$$a_0 = f(1) = 1.$$

$$f'(z) = -1/z^2 = -1 \text{ at } z = 1. \text{ Therefore } a_1 = -1.$$

$$f''(z) = 2/z^3 = 2 \text{ at } z = 1. \text{ Therefore } a_2 = 2/2! = 1.$$

$$f'''(z) = -6/z^4 = -6 \text{ at } z = 1. \text{ Therefore } a_3 = -6/3! = -1.$$

Hence we obtain

$$\frac{1}{z} = 1 - (z-1) + (z-1)^2 - (z-1)^3 + \cdots = \sum_{n=0}^{\infty} (-1)^n (z-1)^n.$$

An alternative method of finding the Taylor expansion of $1/z$ at $z = 1$ is to put $t = z - 1$ and expand in powers of t. We obtain

$$\frac{1}{z} = \frac{1}{1+t} = 1 - t + t^2 - t^3 + \cdots$$
$$= 1 - (z-1) + (z-1)^2 - (z-1)^3 + \cdots$$

as before.

The range of validity for this expansion is $|z-1| < 1$.

3.8 Laurent expansions

The Taylor expansion is a special case of a still more general expansion due to Laurent (1843) which represents $f(z)$ as the sum of a two-way power series

$$f(z) = \sum_{n=-\infty}^{\infty} a_n (z-c)^n$$
$$= \cdots + \frac{a_{-2}}{(z-c)^2} + \frac{a_{-1}}{z-c} + a_0 + a_1(z-c) + a_2(z-c)^2 + \cdots.$$

The Laurent expansion is used for functions which have a singularity at c. We classify singularities according to the type of Laurent expansion obtained. We call that part of the Laurent expansion with negative powers of $z - c$ the *principal part*. We say $f(z)$ has a *pole* at $z = c$ if the principal part has only finitely many non-zero terms. If the principal part has infinitely many non-zero terms we say $f(z)$ has an *essential singularity*.

The *order* of a pole is the largest n for which $a_{-n} \neq 0$. A pole of order 1 is called a *simple* pole. A pole of order 2 is called a *double* pole. The *residue* of $f(z)$ at $z = c$ is the coefficient a_{-1} in the Laurent expansion at $z = c$.

For example, $f(z) = e^{1/z}$ has an essential singularity at $z = 0$, since the Laurent expansion at $z = 0$ is

$$e^{1/z} = 1 + \frac{1}{z} + \frac{1}{2!}\frac{1}{z^2} + \cdots .$$

The residue of $e^{1/z}$ at $z = 0$ is 1.

On the other hand, $g(z) = e^z/z^4$ has a pole of order 4 at $z = 0$, since the Laurent expansion at $z = 0$ is

$$\begin{aligned}\frac{e^z}{z^4} &= \frac{1}{z^4}\left(1 + z + \frac{z^2}{2!} + \frac{z^3}{3!} + \cdots\right)\\ &= \frac{1}{z^4} + \frac{1}{z^3} + \frac{1}{2!}\frac{1}{z^2} + \frac{1}{3!}\frac{1}{z} + \cdots .\end{aligned}$$

The residue of e^z/z^4 at $z = 0$ is $1/3! = 1/6$.

3.9 Calculation of Laurent expansions

We proceed by way of example. Consider the function

$$f(z) = \frac{1}{1+z^2},$$

which has singularities at $z = \pm i$.

We find the Laurent expansion at $z = i$ by putting $t = z - i$ and expanding in powers of t. We obtain

$$\begin{aligned}\frac{1}{1+z^2} &= \frac{1}{1+(t+i)^2} = \frac{1}{1+t^2+2it-1} = \frac{1}{t^2+2it} = \frac{1}{2it}\frac{1}{1+t/2i}\\ &= \frac{1}{2it}\left(1 - \frac{t}{2i} + \left(\frac{t}{2i}\right)^2 - \cdots\right)\\ &= \frac{1}{2it} + \frac{1}{4} - \frac{t}{8i} + \cdots ,\end{aligned}$$

which shows that $f(z)$ has a simple pole at $z = i$ with residue $1/2i$.

We get the Laurent expansion at $z = -i$ by putting $t = z + i$ and expanding in terms of t. This time we have

$$\frac{1}{1+z^2} = -\frac{1}{2it} + \frac{1}{4} + \frac{t}{8i} + \cdots,$$

which shows that $f(z)$ also has a simple pole at $z = -i$, but now with residue $-1/2i$.

Notes

For a proof of Theorem 2, see, for example, Knopp (1945) page 30.

Neither Taylor nor Maclaurin gave a rigorous proof of the validity of their expansions. They are not valid in general, even for functions with derivatives of all orders. An interesting example is the function

$$f(x) = e^{-1/x^2},$$

which (if we assume $f(x) = 0$ at $x = 0$) has $f^{(n)}(0) = 0$ for all n, so has a Maclaurin expansion which vanishes identically, therefore cannot $= f(x)$ at any $x \neq 0$.

They are of course valid for the elementary functions we consider here.

Rigorous treatments of complex analysis are able to give proofs of the validity of Taylor, Maclaurin and Laurent expansions in the complex domain using the theory of contour integration developed in the next chapter. (See Knopp (1945) chapter 7 for the details.)

Examples

1. Verify the Cauchy–Riemann equations for the following functions:

   $$z^3, \quad e^z, \quad \sin z, \quad \log z.$$

   Verify the Cauchy–Riemann formula for the derivative in each case.
2. Prove $|z|^2$ is differentiable only at $z = 0$. What is its derivative at this point?
3. Prove $f(z) = \bar{z}(|z|^2 - 2)$ is differentiable only on the unit circle $|z| = 1$. Verify that $f'(z) = \bar{z}^2$ for these z.
4. Prove that if $f(z)$ is differentiable for all z and is everywhere real valued then $f(z)$ must be constant.
5. Find the Maclaurin expansion of $e^z \sin z$ up to terms in z^5 (i) by differentiating and putting $z = 0$, (ii) by multiplying the Maclaurin expansions of e^z and $\sin z$ together.
6. Find the Taylor expansions of the following functions at the points indicated. State the range of validity in each case.

   (i) $1/z$ at $z = 2$, (ii) e^z at $z = i$, (iii) $\log z$ (PV) at $z = 1$.

7. Find the Laurent expansions of the following functions at the points indicated. State what type of singularity each one is, and what the residues are. Indicate the principal part in each case.

(i) e^z/z^{10} at $z = 0$, (ii) $\sin z/z^{15}$ at $z = 0$, (iii) $\dfrac{1}{z^2 - 1}$ at $z = \pm 1$.

8. Find constants A, B such that

$$f(z) = \frac{3z + 1}{(z + 2)(z - 3)} = \frac{A}{z + 2} + \frac{B}{z - 3}.$$

Hence find the Maclaurin expansion of $f(z)$. What is its range of validity?

Chapter 4

Integrals

4.1 Review of real variables

Suppose that $I = [a, b]$ is a real interval, and that $f(x)$ is a real valued function defined for $x \in I$. Then the integral of $f(x)$ from a to b, or over I, is defined to be

$$\int_a^b f(x)\,dx = \int_I f(x)\,dx = \lim_{dx \to 0} \sum f(x)\,dx.$$

Geometrically the integral represents the area under the graph of $f(x)$ between the limits $x = a$, $x = b$. The approximating area $\sum f(x)\,dx$ represents the sum of the areas of the rectangles height $f(x)$ and width dx (Figure 4.1).

We have the following two theorems.

Theorem 1 (Existence theorem) $f(x)$ continuous implies $f(x)$ integrable.

Theorem 2 (Fundamental theorem of calculus) If $f(x)$ is continuous for $a \leq x \leq b$, then

$$\int_a^b f(x)\,dx = F(b) - F(a) = [F(x)]_a^b,$$

where $F(x)$ is any primitive of $f(x)$.

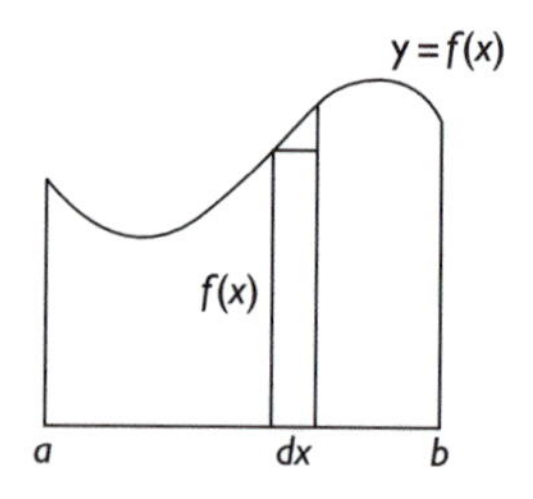

Figure 4.1

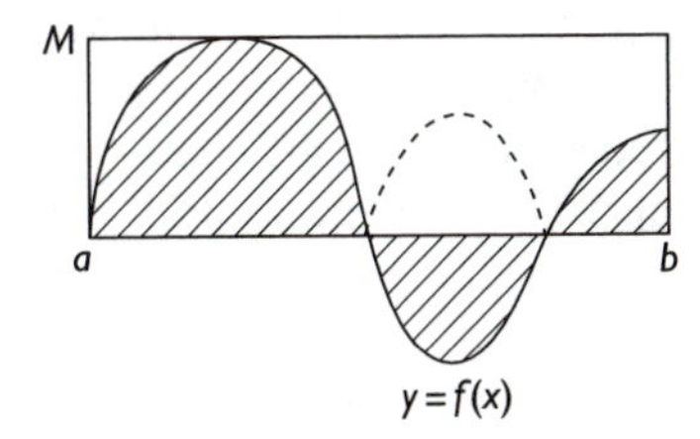

Figure 4.2

An *integrable* function is one for which the above limit exists finite. A *primitive* for $f(x)$ is any $F(x)$ such that $F'(x) = f(x)$. Theorem 1 guarantees that the integral exists for any continuous function. Theorem 2 gives us a practical method for evaluating integrals. Together with the following combination rules.

- Linear combination rule

$$\int_a^b (\lambda f(x) + \mu g(x))\, dx = \lambda \int_a^b f(x) + \mu \int_a^b g(x)\, dx.$$

- Product rule (integration by parts)

$$\int_a^b f(x)g'(x)\, dx = [f(x)g(x)]_a^b - \int_a^b f'(x)g(x)\, dx.$$

- Composite rule (integration by substitution)

$$\int_a^b f(x)\, dx = \int_\alpha^\beta f(g(t))g'(t)\, dt,$$

  where $g(\alpha) = a$, $g(\beta) = b$.

For integrals which cannot be evaluated exactly we have the inequalities

$$\left|\int_a^b f(x)\, dx\right| \le \int_a^b |f(x)|\, dx \le M(b-a),$$

where $|f(x)| \le M$ for $a \le x \le b$ (Figure 4.2).

4.2 Contours

Instead of intervals we shall integrate complex functions $f(z)$ of the complex variable z along contours. By a *contour* γ we mean a continuous curve in the complex plane. A *parametrisation* of γ is a representation of γ as

$$\gamma = \{\phi(t) : \alpha \le t \le \beta\},$$

where $\phi(t)$ is a continuous function on the real interval $[\alpha, \beta]$. We call t the *parameter*, $\phi(t)$ the *parametric function*, $[\alpha, \beta]$ the *parametric interval*. We call the points $a = \phi(\alpha), b = \phi(\beta)$ the *end points* of γ. We say γ is a *closed* contour if $a = b$. The *orientation* of a contour γ is the direction in which the point $z = \phi(t)$ moves as t moves along the parametric interval. We put an arrow on the contour to indicate the orientation.

Example 1 (Straight line) We can parametrise the straight line γ going from a to b as $z = (1-t)a + tb$ where $0 \leq t \leq 1$ (Figure 4.3).

Example 2 (Unit circle) We can parametrise the unit circle γ described once anticlockwise as $z = e^{it}$ where $0 \leq t \leq 2\pi$ (Figure 4.4).

Example 3 (Unit square) The square γ with vertices at $0, 1, 1+i, i$ described once anticlockwise can be written as $\gamma = \gamma_1 + \gamma_2 + \gamma_3 + \gamma_4$, where $\gamma_1, \gamma_2, \gamma_3, \gamma_4$ are the four sides of the square indicated in Figure 4.5. We need a different parametrisation for each side.

On γ_1 We can take $z = t$, where $0 \leq t \leq 1$.
On γ_2 We can take $z = 1 + it$, where $0 \leq t \leq 1$.
On γ_3 We can take $z = t + i$, where $1 \geq t \geq 0$.
On γ_4 We can take $z = it$, where $1 \geq t \geq 0$.

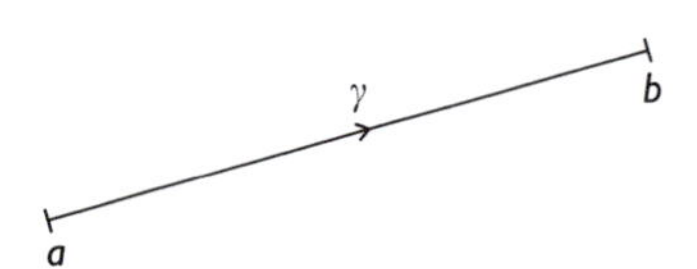

Figure 4.3

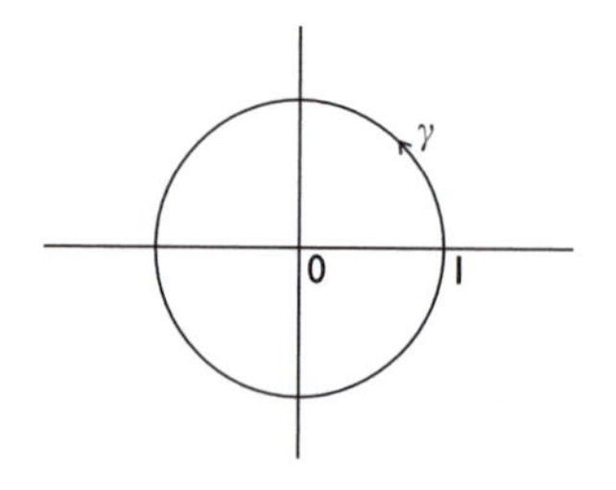

Figure 4.4

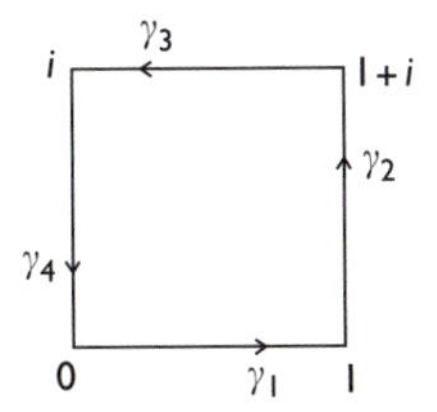

Figure 4.5

Observe that the orientation of γ_3, γ_4 is given by *decreasing* t. We indicate this by writing $1 \geq t \geq 0$ instead of $0 \leq t \leq 1$.

4.3 Contour integrals

Given a contour γ and a function $f(z)$ defined for $z \in \gamma$ we define

$$\int_\gamma f(z)\,dz = \lim_{dz \to 0} \sum f(z)\,dz.$$

Theorems 1 and 2 of Section 4.1 remain valid in the complex context, also the combination rules for integrals. The inequalities generalise to the following.

4.4 Estimate lemma

If $|f(z)| \leq M$ for $z \in \gamma$, then

$$\left|\int_\gamma f(z)\,dz\right| \leq M l_\gamma,$$

where l_γ is the length of γ.

Regarding evaluation of contour integrals we give three methods.

4.5 Method I: Substituting the parametric function

We describe the method by way of examples.

Example 1 Evaluate

$$\int_\gamma z^n\,dz,$$

where γ is the unit circle parametrised by letting $z = e^{it}$ where $0 \leq t \leq 2\pi$.

If we substitute $z = e^{it}$, then we have $dz = ie^{it}dt$. Therefore

$$\begin{aligned}\int_\gamma z^n\,dz &= \int_0^{2\pi} (e^{it})^n ie^{it}\,dt \\ &= i\int_0^{2\pi} e^{i(n+1)t}\,dt \\ &= i\left[\frac{e^{i(n+1)t}}{i(n+1)}\right]_0^{2\pi} \\ &= 0,\end{aligned}$$

if $n \neq -1$. If $n = -1$, then we have

$$\int_\gamma \frac{dz}{z} = i\int_0^{2\pi} dt = 2\pi i.$$

Example 2 Evaluate

$$\int_\gamma z^2\,dz,$$

where γ is the unit square $\gamma = \gamma_1 + \gamma_2 + \gamma_3 + \gamma_4$ as in Figure 4.5.

On γ_1 We have $z = t$ where $0 \leq t \leq 1$. Therefore $dz = dt$ which gives

$$\int_{\gamma_1} z^2\,dz = \int_0^1 t^2\,dt = \frac{1}{3}.$$

On γ_2 We have $z = 1 + it$ where $0 \leq t \leq 1$. Therefore $dz = idt$ which gives

$$\int_{\gamma_2} z^2\,dz = \int_0^1 (1+it)^2 i\,dt = \int_0^1 (i - 2t - it^2)\,dt = -1 + \frac{2}{3}i.$$

On γ_3 We have $z = t + i$ where $1 \geq t \geq 0$. Therefore $dz = dt$ which gives

$$\int_{\gamma_3} z^2\,dz = \int_1^0 (t+i)^2\,dt = \int_1^0 (t^2 + 2it - 1)\,dt = \frac{2}{3} - i.$$

On γ_4 We have $z = it$ where $1 \geq t \geq 0$. Therefore $dz = idt$ which gives

$$\int_{\gamma_4} z^2\,dz = \int_1^0 (it)^2 i\,dt = -i\int_1^0 t^2\,dt = \frac{1}{3}i.$$

Hence we have

$$\int_\gamma z^2\,dz = \int_{\gamma_1} + \int_{\gamma_2} + \int_{\gamma_3} + \int_{\gamma_4} = \frac{1}{3} + \left(-1 + \frac{2}{3}i\right) + \left(\frac{2}{3} - i\right) + \frac{1}{3}i = 0.$$

4.6 Method 2: Using the fundamental theorem of calculus

If the contour γ has end points a, b with orientation a to b, and if the function $f(z)$ has a primitive $F(z)$ on γ ($F'(z) = f(z)$), then

$$\int_\gamma f(z)\,dz = F(b) - F(a) = [F(z)]_a^b.$$

For example, if γ is the parabolic arc $z = t + it^2$ ($0 \le t \le 1$), then

$$\int_\gamma z^2\,dz = \left[\frac{z^3}{3}\right]_0^{1+i} = \frac{(1+i)^3}{3}.$$

Observe that if γ is *any* contour going from 0 to $1+i$ we must have

$$\int_\gamma z^2\,dz = \frac{(1+i)^3}{3}.$$

Observe also that if γ is any *closed* contour ($a = b$), then we must have

$$\int_\gamma z^2\,dz = 0.$$

More generally we have the following theorem.

Theorem 3 If γ is any closed contour, and if $f(z)$ has a primitive on γ then

$$\int_\gamma f(z)\,dz = 0.$$

Corollary 1 (See Example 1 of Section 4.5) If γ is the unit circle, then for all $n \neq -1$ we have

$$\int_\gamma z^n\,dz = 0.$$

Proof For $n \neq -1$ the function z^n has the primitive $z^{n+1}/(n+1)$ on γ.

Corollary 2 The function $1/z$ has no primitive on the unit circle.

Proof We showed in Section 4.5 that

$$\int_\gamma \frac{dz}{z} = 2\pi i \neq 0.$$

It might be thought that $\log z$ is a primitive for $1/z$ on the unit circle. However by Theorem 1 of Chapter 3, any $F(z)$ such that $F'(z) = f(z)$ must be continuous. Whichever values we take for $\log z$ on the unit circle there is bound to be a discontinuity. For example, $\log z$ (PV) has a discontinuity at $z = -1$.

4.7 Method 3: Using the residue theorem

Theorem 3 above says that if $f(z)$ has a primitive on the closed contour γ, then

$$\int_\gamma f(z)\,dz = 0.$$

The analogue of this theorem for derivatives is as follows.

Theorem 4 (Cauchy's theorem) If γ is a closed contour and if $f(z)$ has a derivative on γ and everywhere inside γ, then

$$\int_\gamma f(z)\,dz = 0.$$

Proof See Appendix 1.

We can use Cauchy's theorem to show (for a third time) that if γ is the unit circle, then

$$\int_\gamma z^n\,dz = 0$$

for $n \geq 0$. For $n < 0$ Cauchy's theorem does not tell us anything, since z^n then has a singularity at $z = 0$ which is inside γ.

Cauchy's theorem might appear at first sight to be rather trivial. However, it turns out to have far reaching consequences as we shall shortly see.

Corollary 1 If the contours γ_1, γ_2 have the same end points a, b and if $f(z)$ is differentiable on γ_1, γ_2 and between them, then (Figure 4.6)

$$\int_{\gamma_1} f(z)\,dz = \int_{\gamma_2} f(z)\,dz.$$

Proof If $\gamma = \gamma_2 - \gamma_1$, then we can apply Cauchy's theorem to γ to obtain

$$0 = \int_\gamma f(z)\,dz = \int_{\gamma_2} f(z)\,dz - \int_{\gamma_1} f(z)\,dz.$$

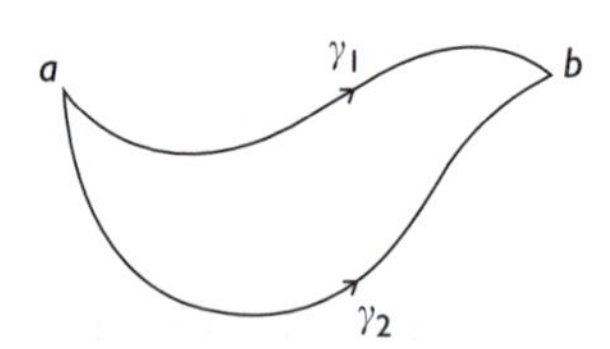

Figure 4.6

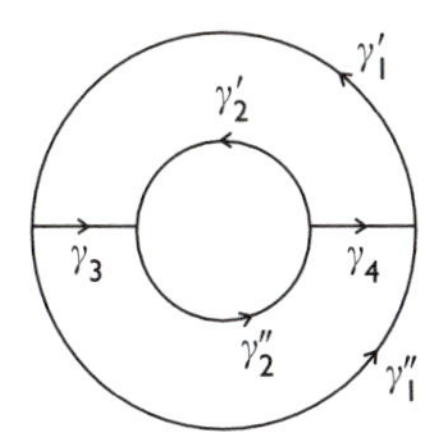

Figure 4.7

Corollary 2 If the closed contours γ_1, γ_2 are such that γ_2 lies inside γ_1, and if $f(z)$ is differentiable on γ_1, γ_2 and between them, then

$$\int_{\gamma_1} f(z)\,dz = \int_{\gamma_2} f(z)\,dz.$$

Proof If we make cross cuts γ_3, γ_4 as indicated in Figure 4.7, and if we denote the upper parts of γ_1, γ_2 by γ_1', γ_2' and the lower parts by γ_1'', γ_2'' then by Corollary 1 we have

$$\int_{\gamma_1'} f(z)\,dz = \int_{\gamma_2'} f(z)\,dz - \int_{\gamma_3} f(z)\,dz - \int_{\gamma_4} f(z)\,dz,$$

$$\int_{\gamma_1''} f(z)\,dz = \int_{\gamma_2''} f(z)\,dz + \int_{\gamma_3} f(z)\,dz + \int_{\gamma_4} f(z)\,dz.$$

Therefore

$$\begin{aligned}\int_{\gamma_1} f(z)\,dz &= \int_{\gamma_1'} f(z)\,dz + \int_{\gamma_1''} f(z)\,dz \\ &= \int_{\gamma_2'} f(z)\,dz + \int_{\gamma_2''} f(z)\,dz = \int_{\gamma_2} f(z)\,dz.\end{aligned}$$

Corollary 3 If non-intersecting closed contours $\gamma_1, \ldots, \gamma_n$ all lie inside the closed contour γ, and if $f(z)$ is differentiable on $\gamma, \gamma_1, \ldots, \gamma_n$ and on the area internal to γ and external to $\gamma_1, \ldots, \gamma_n$, then (Figure 4.8)

$$\int_{\gamma} f(z)\,dz = \sum_{k=1}^{n} \int_{\gamma_k} f(z)\,dz.$$

Proof Make cross cuts as in the proof of Corollary 2.

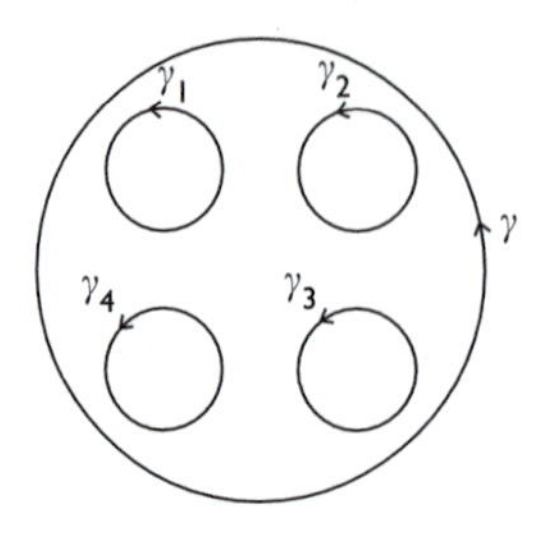

Figure 4.8

Theorem 5 (Residue theorem) If γ is a closed contour and if $f(z)$ is differentiable on γ and inside γ except at $c_1, \ldots, c_n$ inside γ, then

$$\int_\gamma f(z)\,dz = 2\pi i \sum_{k=1}^{n} R_k,$$

where R_k is the residue of $f(z)$ at c_k.

Proof (Special case) Suppose $f(z)$ has a single singularity at $z = c$ inside γ. If we let γ_r be a circle centre c, radius r small enough to ensure that γ_r lies inside γ, then by Corollary 2 of Cauchy's theorem we have

$$\int_\gamma f(z)\,dz = \int_{\gamma_r} f(z)\,dz.$$

If the Laurent expansion of $f(z)$ at $z = c$ is

$$f(z) = \sum_{-\infty}^{\infty} a_n (z-c)^n,$$

then we have

$$\int_{\gamma_r} f(z)\,dz = \sum_{-\infty}^{\infty} a_n \int_{\gamma_r} (z-c)^n\,dz = a_{-1} \int_{\gamma_r} \frac{dz}{z-c}$$

since for $n \neq -1$ we have

$$\int_{\gamma_r} (z-c)^n\,dz = \left[\frac{(z-c)^{n+1}}{n+1}\right]_{\gamma_r} = 0$$

by Method 2.

On γ_r we have $z = c + re^{it}$ $(0 \le t \le 2\pi)$, therefore by Method 1 we have

$$\int_{\gamma_r} \frac{dz}{z-c} = \int_0^{2\pi} \frac{ire^{it}}{re^{it}}\, dt = \int_0^{2\pi} i\, dt = 2\pi i.$$

Hence

$$\int_\gamma f(z)\, dz = 2\pi i a_{-1}$$

as required.

General case If $f(z)$ has a singularities at $z = c_1, \ldots, c_n$ inside γ then we can draw circles $\gamma_1, \ldots, \gamma_n$ with centres at $c_1, \ldots, c_n$ and with radii small enough to ensure they all lie inside γ and that they don't intersect each other. Hence by Corollary 3 of Cauchy's theorem we have

$$\int_\gamma f(z)\, dz = \sum_{k=1}^{n} \int_{\gamma_k} f(z)\, dz = \sum_{k=1}^{n} 2\pi i R_k$$

by what we have already proved.

Example Evaluate the integral

$$\int_\gamma \frac{dz}{z^2+1},$$

where γ is to be specified.

Answer We need to find the singularities of the integrand, and find the residues at these singularities. In fact we already did this in Section 3.9 where we found that the singularities are at $z = \pm i$ with residues $\pm 1/2i$.

γ = *circle centre i, radius* 1. In this case i is inside γ, $-i$ is outside γ. Therefore,

$$\int_\gamma \frac{dz}{z^2+1} = 2\pi i \left(\frac{1}{2i}\right) = \pi.$$

γ = *circle centre* $-i$, *radius* 1. In this case $-i$ is inside γ, i is outside γ. Therefore,

$$\int_\gamma \frac{dz}{z^2+1} = 2\pi i \left(-\frac{1}{2i}\right) = -\pi.$$

γ = *circle centre 0, radius* 2. In this case both singularities $\pm i$ are inside γ. Therefore,

$$\int_\gamma \frac{dz}{z^2+1} = 2\pi i \left(\frac{1}{2i} - \frac{1}{2i}\right) = 0.$$

γ = *circle centre 0, radius* 1/2. In this case neither singularity is inside γ. Therefore,

$$\int_\gamma \frac{dz}{z^2+1} = 0$$

by Cauchy's theorem.

4.8 Quick ways of finding residues

For simple poles there are quicker methods for finding residues than calculating the Laurent expansion and taking the -1th Laurent coefficient. For example, we have the following.

Cover up rule If $f(z)$ takes the form

$$f(z) = \frac{g(z)}{z-c},$$

where $g(c) \neq 0$, then the residue of $f(z)$ at $z = c$ is $g(c)$.

Proof The Taylor expansion for $g(z)$ at $z = c$ is

$$g(z) = g(c) + (z-c)g'(c) + \cdots,$$

which gives immediately

$$f(z) = \frac{g(c)}{z-c} + g'(c) + \cdots$$

Example Consider again

$$\frac{1}{z^2+1} = \frac{1}{(z+i)(z-i)},$$

which has simple poles at $z = \pm i$. Covering up $z - i$, $z + i$ in turn we have

$$\operatorname*{Res}_{z=i} \frac{1}{z^2+1} = \left[\frac{1}{z+i}\right]_{z=i} = \frac{1}{2i}, \quad \operatorname*{Res}_{z=-i} \frac{1}{z^2+1} = \left[\frac{1}{z-i}\right]_{z=-i} = -\frac{1}{2i}.$$

Differentiating the denominator If $f(z)$ takes the form $f(z) = g(z)/h(z)$, where $g(c) \neq 0$, $h(c) = 0$, $h'(c) \neq 0$ then the residue of $f(z)$ at $z = c$ is $g(c)/h'(c)$.

Proof We have

$$f(z) = \frac{g(c) + (z-c)g'(c) + \cdots}{(z-c)h'(c) + (z-c)^2h''(c)/2! + \cdots} = \frac{k(z)}{z-c},$$

where

$$k(z) = \frac{g(c) + (z-c)g'(c) + \cdots}{h'(c) + (z-c)h''(c)/2! + \cdots}.$$

Therefore the residue of $f(z)$ at $z = c$ is $g(c)/h'(c)$ by the cover up rule.

Example Consider again

$$\operatorname*{Res}_{z=i} \frac{1}{z^2+1} = \left[\frac{1}{2z}\right]_{z=i} = \frac{1}{2i}, \quad \operatorname*{Res}_{z=-i} \frac{1}{z^2+1} = \left[\frac{1}{2z}\right]_{z=-i} = -\frac{1}{2i}.$$

Notes

A rigorous treatment of contour integration would present the facts in a different order. We have assumed in our proof of the residue theorem that a differentiable function has a valid Laurent expansion near an isolated singularity, and that this expansion can be integrated term by term. We have also assumed in our statement of Cauchy's theorem that the 'inside' of a closed contour is well defined. A rigorous proof of the residue theorem requires a knowledge of uniform convergence. A rigorous proof of Cauchy's theorem requires a knowledge of plane topology. Both of these can be found in Knopp (1945).

Examples

1. Evaluate the following contour integrals.

   (i) $\int_\gamma \operatorname{Re} z\, dz$ where γ is the unit circle $z = e^{it}$ $(0 \leq t \leq 2\pi)$.

   (ii) $\int_\gamma |z|^2\, dz$ where γ is the parabolic arc $z = t + it^2$ $(0 \leq t \leq 1)$.

   (iii) $\int_\gamma \bar{z}\, dz$ where γ is the straight line joining 0 to $1 + i$.

2. Use the estimate lemma to prove the following inequalities.

   (i) $\left|\int_\gamma \frac{e^z}{z-1}\, dz\right| \leq 2\pi e^2$ where γ is the circle $|z-1| = 1$.

   (ii) $\left|\int_\gamma \frac{\sin z}{z+i}\, dz\right| \leq \frac{\pi \sinh 1}{\sqrt{2}}$ where γ is the semicircle $z = e^{it}$ $(0 \leq t \leq \pi)$.

(iii) $\left|\int_\gamma \frac{z-2}{z-3}\,dz\right| \leq 4\sqrt{10}$ where γ is the square with vertices at $\pm 1 \pm i$.

3. Find all the singularities of the following functions. Use the method of differentiating the denominator to find all the residues.

$$\text{(i)}\ \frac{z+1}{z-1}, \quad \text{(ii)}\ \frac{e^z}{z^2+\pi^2}, \quad \text{(iii)}\ \frac{1}{z^2-6z+8}.$$

4. Use the residue theorem to evaluate the following integrals round the contours indicated.

(i) $\displaystyle\int_\gamma \frac{z+1}{z-1}\,dz$ (γ = circle centre 1, radius 1).

(ii) $\displaystyle\int_\gamma \frac{e^z}{z^2+\pi^2}\,dz$ (γ = circle centre πi, radius π).

(iii) $\displaystyle\int_\gamma \frac{dz}{z^2-6z+8}$ (γ = circle centre 0, radius 3).

5. Prove that if $f(z)$ is differentiable inside and on the closed contour γ, then for any a inside γ

$$f(a) = \frac{1}{2\pi i}\int_\gamma \frac{f(z)\,dz}{z-a}$$

(Cauchy's integral formula). What is the value of this integral if a is outside γ?

Chapter 5

Evaluation of finite real integrals

As a first application of the residue theorem (see Section 4.7) we describe a method for evaluating a certain class of real integrals over a finite interval.

Example 1 Consider the integral

$$\int_0^{2\pi} \frac{dt}{5 + 4\cos t}.$$

We can transform this integral into a contour integral round the unit circle by making the substitution $z = e^{it}$. We have $dz = ie^{it}dt = izdt$ which gives $dt = dz/iz$. We also have

$$\cos t = \frac{e^{it} + e^{-it}}{2} = \frac{z + 1/z}{2}.$$

Therefore we get

$$\int_0^{2\pi} \frac{dt}{5 + 4\cos t} = \int_\gamma \frac{dz}{iz} \frac{1}{5 + 2(z + 1/z)} = \frac{1}{i} \int_\gamma \frac{dz}{2z^2 + 5z + 2},$$

where γ is the unit circle.

We now evaluate this contour integral using the residue theorem. Observe that $2z^2 + 5z + 2 = (2z + 1)(z + 2)$, therefore the singularities of the integrand occur at $z = -2, -1/2$ (Figure 5.1). Of these only $z = -1/2$ is inside γ, where the residue is

$$\frac{1}{i}\left[\frac{1}{4z + 5}\right]_{z=-1/2} = \frac{1}{3i},$$

using the method of differentiating the denominator (see Section 4.8).

Hence we have

$$\int_0^{2\pi} \frac{dt}{5 + 4\cos t} = \frac{1}{i} \int_\gamma \frac{dz}{2z^2 + 5z + 2} = 2\pi i \times \frac{1}{3i} = \frac{2\pi}{3}.$$

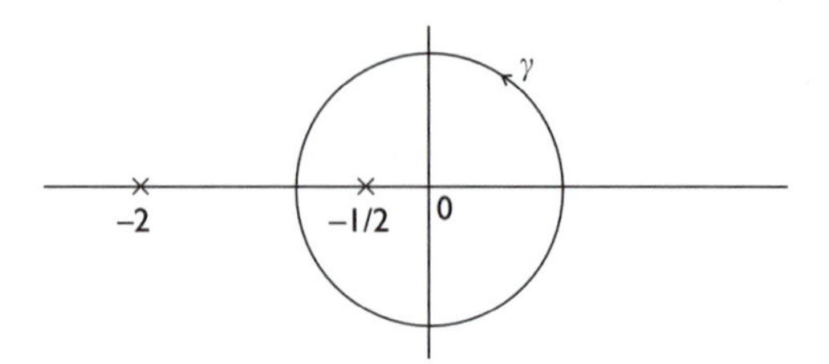

Figure 5.1

Example 2 Consider the integral

$$\int_0^{2\pi} \frac{dt}{1+\sin^2 t}.$$

Putting $z = e^{it}$ gives

$$\int_0^{2\pi} \frac{dt}{1+\sin^2 t} = \int_\gamma \frac{dz}{iz} \frac{1}{1-(z-1/z)^2/4} = \int_\gamma \frac{4iz\,dz}{z^4-6z^2+1},$$

where γ is the unit circle. Here we used the formula

$$\sin t = \frac{e^{it}-e^{-it}}{2i} = \frac{z-1/z}{2i}.$$

The singularities of the integrand are at the solutions of the equation

$$z^4 - 6z^2 + 1 = 0,$$

which are given by

$$z^2 = \frac{6 \pm \sqrt{36-4}}{2} = 3 \pm 2\sqrt{2}.$$

Of these only $z = \pm\sqrt{3-2\sqrt{2}} = \pm(\sqrt{2}-1)$ are inside γ (Figure 5.2). Differentiating the denominator we obtain the residues by evaluating

$$\frac{4iz}{4z^3-12z} = \frac{i}{z^2-3}$$

at $z^2 = 3 - 2\sqrt{2}$. Therefore the residues are $-i/2\sqrt{2}$ at both these points.

Hence by the residue theorem we have

$$\int_0^{2\pi} \frac{dt}{1+\sin^2 t} = \int_\gamma \frac{4iz\,dz}{z^4-6z^2+1} = -\frac{i}{2\sqrt{2}} \times 2 \times 2\pi i = \pi\sqrt{2}.$$

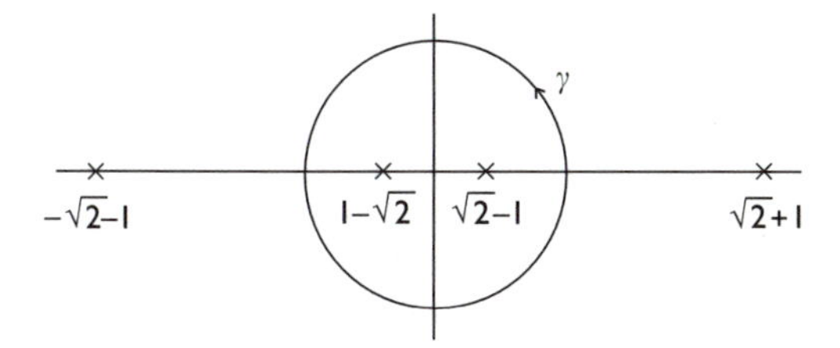

Figure 5.2

Example 3 Consider the integral

$$\int_0^{2\pi} \cos^4 t\, dt.$$

The substitution $z = e^{it}$ gives

$$\begin{aligned}\int_0^{2\pi} \cos^4 t\, dt &= \int_\gamma \frac{dz}{16iz}\left(z + \frac{1}{z}\right)^4 \\ &= \int_\gamma \frac{dz}{16iz}\left(z^4 + 4z^2 + 6 + \frac{4}{z^2} + \frac{1}{z^4}\right) \\ &= \frac{1}{16i}\int_\gamma \left(z^3 + 4z + \frac{6}{z} + \frac{4}{z^3} + \frac{1}{z^5}\right) dz,\end{aligned}$$

where γ is the unit circle.

Observe that the integrand is already a Laurent expansion, indicating that there is a pole of order 5 at $z = 0$, and that the residue there is $6/16i = 3/8i$.

Hence we have

$$\int_0^{2\pi} \cos^4 t\, dt = 2\pi i \times \frac{3}{8i} = \frac{3\pi}{4}.$$

Example 4 Consider the integral

$$\int_0^{2\pi} \sin 2t \cos 3t\, dt.$$

Making the substitution $z = e^{it}$ we obtain

$$\begin{aligned}\int_0^{2\pi} \sin 2t \cos 3t \, dt &= \frac{1}{4i} \int_\gamma \frac{dz}{iz} \left(z^2 - \frac{1}{z^2}\right)\left(z^3 + \frac{1}{z^3}\right) \\ &= \frac{1}{4i} \int_\gamma \frac{dz}{iz} \left(z^5 - z + \frac{1}{z} - \frac{1}{z^5}\right) \\ &= -\frac{1}{4} \int_\gamma \left(z^4 - 1 + \frac{1}{z^2} - \frac{1}{z^6}\right) dz \\ &= 0,\end{aligned}$$

since the integrand is a Laurent expansion with no term in $1/z$.

Hence we have

$$\int_0^{2\pi} \sin 2t \cos 3t \, dt = 0.$$

Examples

Evaluate the following integrals.

1. $\displaystyle\int_0^{2\pi} \frac{dt}{2 + \cos t} \quad (2\pi/\sqrt{3})$
2. $\displaystyle\int_0^{2\pi} \frac{dt}{3 + 2\sin t} \quad (2\pi/\sqrt{5})$
3. $\displaystyle\int_0^{2\pi} \frac{dt}{4 - 3\cos^2 t} \quad (\pi)$
4. $\displaystyle\int_0^{2\pi} \frac{\sin 5t}{\sin t} \, dt \quad (2\pi)$
5. $\displaystyle\int_0^{2\pi} \cos^6 t \, dt \quad (5\pi/8)$

Chapter 6

Evaluation of infinite real integrals

6.1 Convergence

For integrals of the form

$$\int_{-\infty}^{\infty} f(x)\,dx$$

the problem of convergence arises. We shall define the *Cauchy principal value* (CPV) of such an integral to be

$$\int_{-\infty}^{\infty} f(x)\,dx = \lim_{R\to\infty} \int_{-R}^{R} f(x)\,dx,$$

and say the integral *converges* whenever this limit exists. For example, consider the integral

$$\int_{-\infty}^{\infty} \frac{dx}{x^2+1}.$$

We have

$$\int_{-R}^{R} \frac{dx}{x^2+1} = \left[\tan^{-1} x\right]_{-R}^{R} = 2\tan^{-1} R \to 2 \times \frac{\pi}{2} = \pi$$

as $R \to \infty$. Therefore, the integral converges and its value is

$$\int_{-\infty}^{\infty} \frac{dx}{x^2+1} = \pi.$$

6.2 The method

We illustrate the method for evaluating infinite real integrals using complex calculus by applying it to the integral

$$\int_{-\infty}^{\infty} \frac{dx}{x^2+1}$$

considered in Section 6.1.

Let $\gamma = \gamma_1 + \gamma_2$ be the D-shaped contour consisting of the real interval $[-R, R] = \gamma_1$ together with the upper semicircle γ_2 having $[-R, R]$ as diameter (Figure 6.1). And consider the contour integral

$$\int_{\gamma} \frac{dz}{z^2+1} = \int_{\gamma_1} \frac{dz}{z^2+1} + \int_{\gamma_2} \frac{dz}{z^2+1}.$$

On γ_1 We have $z = x$ $(-R \leq x \leq R)$ therefore

$$\int_{\gamma_1} \frac{dz}{z^2+1} = \int_{-R}^{R} \frac{dx}{x^2+1}.$$

Inside γ For all $R > 1$ the integrand has one singularity at $z = i$ where the residue is $1/2i$. Therefore

$$\int_{\gamma} \frac{dz}{z^2+1} = 2\pi i \times \frac{1}{2i} = \pi.$$

On γ_2 We have $|z| = R$ therefore

$$|z^2+1| \geq R^2 - 1$$

(see Inequality 5 of Section 1.11) from which it follows that

$$\left|\frac{1}{z^2+1}\right| \leq \frac{1}{R^2-1}$$

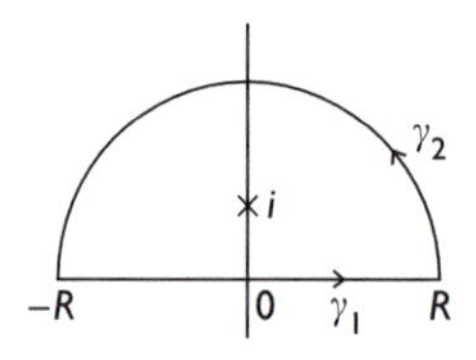

Figure 6.1

provided $R > 1$. We also have $l_\gamma = \pi R$. Therefore by the estimate lemma (see Section 4.4) we obtain

$$\left| \int_{\gamma_2} \frac{dz}{z^2+1} \right| \leq \frac{\pi R}{R^2 - 1} \to 0$$

as $R \to \infty$.

Putting all this information together we have

$$\int_{-R}^{R} \frac{dx}{x^2+1} = \int_{\gamma_1} \frac{dz}{z^2+1} = \int_{\gamma} \frac{dz}{z^2+1} - \int_{\gamma_2} \frac{dz}{z^2+1} = \pi - \int_{\gamma_2} \frac{dz}{z^2+1} \to \pi$$

as $R \to \infty$. Hence we deduce that

$$\int_{-\infty}^{\infty} \frac{dx}{x^2+1} \text{converges} = \pi$$

agreeing with what we found in Section 6.1.

6.3 Failure of $\int_{\gamma_2} \to 0$

It is essential to the success of the method outlined in Section 6.2 that we can prove $\int_{\gamma_2} \to 0$. For example, consider the integral

$$\int_{-\infty}^{\infty} \frac{x\,dx}{x^2+1}.$$

In this case, we have

$$\operatorname*{Res}_{z=i} \frac{z}{z^2+1} = \left[\frac{z}{2z}\right]_{z=i} = \frac{1}{2}.$$

Therefore,

$$\int_{\gamma} \frac{z\,dz}{z^2+1} = 2\pi i \times \frac{1}{2} = \pi i$$

for $R > 1$. If we could prove that

$$\int_{\gamma_2} \frac{z\,dz}{z^2+1} \to 0$$

as $R \to \infty$, then we could deduce that

$$\int_{-R}^{R} \frac{x\,dx}{x^2+1} = \int_{\gamma_1} \frac{z\,dz}{z^2+1} = \int_{\gamma} \frac{z\,dz}{z^2+1} - \int_{\gamma_2} \frac{z\,dz}{z^2+1} \to \pi i$$

as $R \to \infty$. Which is nonsense because

$$\int_{-R}^{R} \frac{x\,dx}{x^2+1} = 0$$

for all R since the integrand is odd. In fact

$$\int_{\gamma_2} \frac{z\,dz}{z^2+1} \not\to 0$$

in this case. Its value is

$$\int_{\gamma_2} \frac{z\,dz}{z^2+1} = \int_{\gamma} \frac{z\,dz}{z^2+1} - \int_{\gamma_1} \frac{z\,dz}{z^2+1} = \pi i$$

for all $R > 1$.

6.4 Integrals involving $\cos x$, $\sin x$

Consider the integral

$$\int_{-\infty}^{\infty} \frac{\cos x\,dx}{x^2+1}.$$

The contour integral

$$\int_{\gamma} \frac{\cos z\,dz}{z^2+1},$$

where $\gamma = \gamma_1 + \gamma_2$ as in Section 6.2 will be no use here because

$$|\cos z|^2 = \cos^2 x + \sinh^2 y$$

($z = x + iy$) is unbounded in the upper half plane. Instead we use

$$\int_{\gamma} \frac{e^{iz}\,dz}{z^2+1},$$

observing first that the integral we require

$$\int_{-\infty}^{\infty} \frac{\cos x\,dx}{x^2+1} = \text{Re} \int_{-\infty}^{\infty} \frac{e^{ix}\,dx}{x^2+1},$$

and second that e^{iz} is bounded in the upper half plane since

$$|e^{iz}| = |e^{i(x+iy)}| = |e^{ix-y}| = |e^{ix}|\,|e^{-y}| = e^{-y} \le 1$$

for $y \geq 0$. It follows that

$$\left| \int_{\gamma_2} \frac{e^{iz}\,dz}{z^2+1} \right| \leq \frac{\pi R}{R^2-1} \to 0$$

as $R \to \infty$. For $R > 1$ the integrand has a singularity at $z = i$ inside γ where the residue is

$$\operatorname*{Res}_{z=i} \frac{e^{iz}}{z^2+1} = \left[\frac{e^{iz}}{2z} \right]_{z=i} = \frac{1}{2ie}.$$

Therefore,

$$\int_{\gamma} \frac{e^{iz}\,dz}{z^2+1} = 2\pi i \times \frac{1}{2ie} = \frac{\pi}{e}.$$

Hence we have

$$\int_{-R}^{R} \frac{e^{ix}\,dx}{x^2+1} = \int_{\gamma_1} \frac{e^{iz}\,dz}{z^2+1} = \int_{\gamma} \frac{e^{iz}\,dz}{z^2+1} - \int_{\gamma_2} \frac{e^{iz}\,dz}{z^2+1} = \frac{\pi}{e} - \int_{\gamma_2} \frac{e^{iz}\,dz}{z^2+1} \to \frac{\pi}{e}$$

as $R \to \infty$, from which it follows that

$$\int_{-\infty}^{\infty} \frac{\cos x\,dx}{x^2+1} \text{ converges } = \frac{\pi}{e}.$$

We therefore deduce that

$$\int_{-\infty}^{\infty} \frac{\cos x\,dx}{x^2+1} = \operatorname{Re} \int_{-\infty}^{\infty} \frac{e^{ix}\,dx}{x^2+1} = \frac{\pi}{e}.$$

We can also deduce that

$$\int_{-\infty}^{\infty} \frac{\sin x\,dx}{x^2+1} = \operatorname{Im} \int_{-\infty}^{\infty} \frac{e^{ix}\,dx}{x^2+1} = 0,$$

though this is of course immediate from the fact that the integrand is odd in this case.

6.5 Roots of unity

Suppose we want to evaluate the integral

$$\int_{-\infty}^{\infty} \frac{dx}{x^4+1}$$

by considering the contour integral

$$\int_{\gamma} \frac{dz}{z^4+1},$$

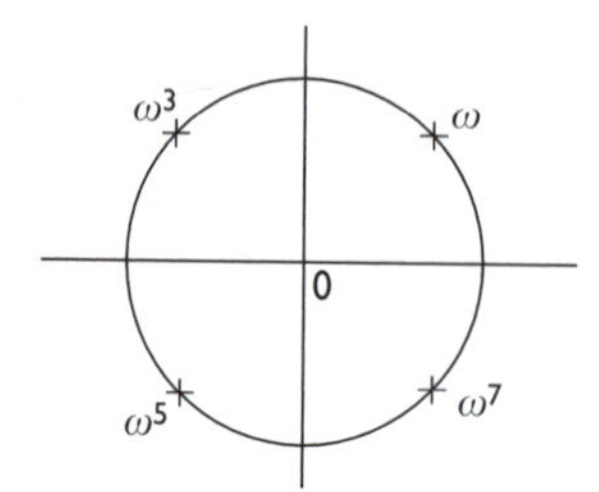

Figure 6.2

where $\gamma = \gamma_1 + \gamma_2$ as in Section 6.2. The integrand has singularities at the solutions of the equation $z^4 + 1 = 0$ which are $\omega, \omega^3, \omega^5, \omega^7$ (Figure 6.2) where $\omega = e^{i\pi/4}$ is the primitive 8th root of unity (see Section 1.10).

Differentiating the denominator we find the residues of the integrand at $z = \omega, \omega^3$ are $1/4\omega^3, 1/4\omega^9$, respectively. Therefore,

$$\int_\gamma \frac{dz}{z^4+1} = 2\pi i\left(\frac{1}{4\omega^3} + \frac{1}{4\omega^9}\right) = \frac{\pi i}{2\omega^9}(\omega^6+1) = \frac{\pi i(1-i)}{2\omega} = \frac{\pi(1+i)}{2\omega}$$

since $\omega^6 = -i$. But $\omega = (1+i)/\sqrt{2}$ so we get

$$\int_\gamma \frac{dz}{z^4+1} = \frac{\pi}{\sqrt{2}}.$$

By the estimate lemma (4.4) we have

$$\left|\int_{\gamma_2} \frac{dz}{z^4+1}\right| \leq \frac{\pi R}{R^4-1} \to 0$$

as $R \to \infty$. Therefore,

$$\int_{-R}^{R} \frac{dx}{x^4+1} = \int_{\gamma_1} \frac{dz}{z^4+1} = \int_\gamma \frac{dz}{z^4+1} - \int_{\gamma_2} \frac{dz}{z^4+1} \to \frac{\pi}{\sqrt{2}}$$

as $R \to \infty$, which shows that

$$\int_{-\infty}^{\infty} \frac{dx}{x^4+1} \text{ converges } = \frac{\pi}{\sqrt{2}}.$$

6.6 Singularities on the real axis

We cannot evaluate the integral

$$\int_0^{\infty} \frac{dx}{x^3+1}$$

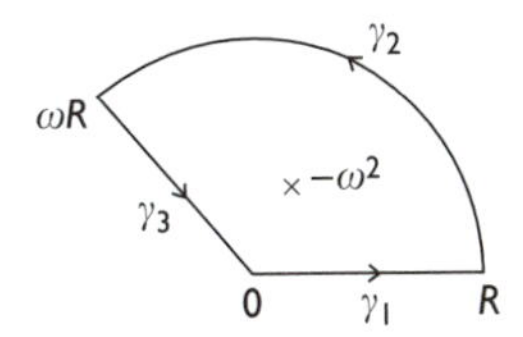

Figure 6.3

by considering the contour integral

$$\int_\gamma \frac{dz}{z^3+1}$$

round γ as in Section 6.2 since the integrand has a singularity at $z = -1$ which is on γ. Instead we use the pizza slice contour $\gamma = \gamma_1 + \gamma_2 + \gamma_3$ shown in Figure 6.3. Here $\omega = e^{2\pi i/3}$ is the primitive cube root of unity (see Section 1.10).

On γ_1 We have $z = t$ $(0 \le t \le R)$. Therefore,

$$\int_{\gamma_1} \frac{dz}{z^3+1} = \int_0^R \frac{dt}{t^3+1}.$$

On γ_3 We have $z = \omega t$ $(R \ge t \ge 0)$. Therefore,

$$\int_{\gamma_3} \frac{dz}{z^3+1} = -\int_0^R \frac{\omega\, dt}{\omega^3 t^3 + 1} = -\omega \int_0^R \frac{dt}{t^3+1} \quad (\omega^3 = 1).$$

On γ_2 We have

$$\left| \int_{\gamma_2} \frac{dz}{z^3+1} \right| \le \frac{2\pi R/3}{R^3 - 1} \to 0$$

as $R \to \infty$.

The integrand has a singularity inside γ at $z = e^{i\pi/3} = -\omega^2$ (if $R > 1$) where the residue is

$$\operatorname*{Res}_{z=-\omega^2} \frac{1}{z^3+1} = \left[\frac{1}{3z^2}\right]_{z=-\omega^2} = \frac{1}{3\omega^4} = \frac{1}{3\omega}.$$

Therefore,

$$\int_\gamma \frac{dz}{z^3+1} = \frac{2\pi i}{3\omega}.$$

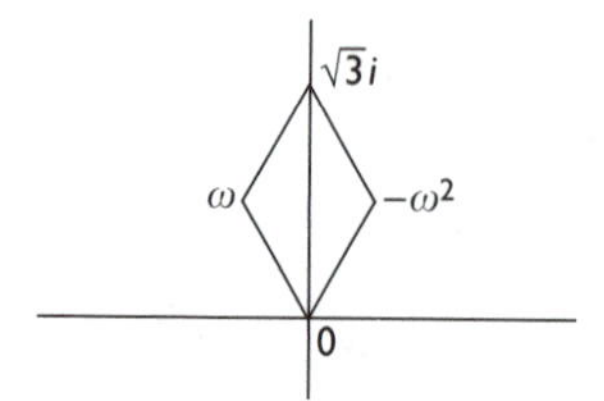

Figure 6.4

Hence we have

$$\frac{2\pi i}{3\omega} = (1-\omega)\int_0^R \frac{dt}{t^3+1} + \int_{\gamma_2} \frac{dz}{z^3+1},$$

which shows that

$$\int_0^\infty \frac{dt}{t^3+1} \text{ converges} = \frac{2\pi i}{3\omega(1-\omega)} = \frac{2\pi}{3\sqrt{3}},$$

since $\omega(1-\omega) = \omega - \omega^2 = \sqrt{3}i$ (see Figure 6.4).

6.7 Half residue theorem

To evaluate the integral

$$\int_{-\infty}^\infty \frac{\sin x}{x}\, dx$$

we would like to consider the contour integral

$$\int_\gamma \frac{e^{iz}}{z}\, dz$$

round the usual D-shaped contour as in Section 6.4 except that the integrand has a singularity at $z = 0$. We use instead the indented contour $\gamma = \gamma_1 + \gamma_2 + \gamma_3 + \gamma_4$ shown in Figure 6.5.

On γ_1 We have $z = t$ $(r \le t \le R)$. Therefore,

$$\int_{\gamma_1} \frac{e^{iz}}{z}\, dz = \int_r^R \frac{e^{it}}{t}\, dt.$$

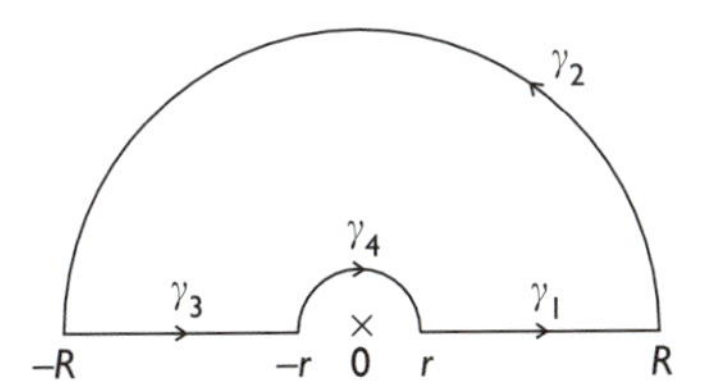

Figure 6.5

On γ_3 We have $z = -t$ $(R \geq t \geq r)$. Therefore,

$$\int_{\gamma_3} \frac{e^{iz}}{z}\, dz = -\int_r^R \frac{e^{-it}}{t}\, dt.$$

Combining these two integrals we get

$$\int_{\gamma_1} \frac{e^{iz}}{z}\, dz + \int_{\gamma_3} \frac{e^{iz}}{z}\, dz = \int_r^R \frac{e^{it} - e^{-it}}{t}\, dt = 2i \int_r^R \frac{\sin t}{t}\, dt.$$

On γ_2 We have (integrating by parts)

$$\int_{\gamma_2} \frac{e^{iz}}{z}\, dz = \left[\frac{e^{iz}}{iz}\right]_{\gamma_2} + \int_{\gamma_2} \frac{e^{iz}}{iz^2}\, dz.$$

The first term on the right-hand side

$$\left[\frac{e^{iz}}{iz}\right]_{\gamma_2} = \frac{e^{-iR}}{-iR} - \frac{e^{iR}}{iR} = -\frac{2\cos R}{iR} \to 0$$

as $R \to \infty$. Whilst the second term

$$\left|\int_{\gamma_2} \frac{e^{iz}}{iz^2}\, dz\right| \leq \frac{\pi R}{R^2} = \frac{\pi}{R} \to 0$$

as $R \to \infty$. Therefore,

$$\int_{\gamma_2} \frac{e^{iz}}{z}\, dz \to 0$$

as $R \to \infty$.

On γ_4 We have $z = re^{it}$ $(\pi \geq t \geq 0)$. Therefore,

$$\int_{\gamma_4} \frac{e^{iz}}{z}\, dz = \int_{\gamma_4} \frac{dz}{z} + \int_{\gamma_4} \frac{1}{z} \sum_1^\infty \frac{(iz)^n}{n!}\, dz.$$

The first term on the right-hand side

$$\int_{\gamma_4} \frac{dz}{z} = -\int_0^{\pi} \frac{ire^{it}}{re^{it}}\, dt = -\int_0^{\pi} i\, dt = -\pi i.$$

As for the second term, we have

$$\left| \frac{1}{z} \sum_1^{\infty} \frac{(iz)^n}{n!} \right| \le \frac{1}{r} \sum_1^{\infty} \frac{r^n}{n!} = \frac{e^r - 1}{r},$$

which gives

$$\left| \int_{\gamma_4} \frac{1}{z} \sum_1^{\infty} \frac{(iz)^n}{n!}\, dz \right| \le \pi r \left(\frac{e^r - 1}{r} \right) = \pi(e^r - 1) \to 0$$

as $r \to 0$. Therefore,

$$\int_{\gamma_4} \frac{e^{iz}}{z}\, dz \to -\pi i$$

as $r \to 0$.

This is a particular case of the half residue theorem. If γ were a full circle centre 0, radius r, then

$$\int_{\gamma_4} \frac{e^{iz}}{z}\, dz = 2\pi i$$

for all $r > 0$, since the residue of the integrand at $z = 0$ is 1. The half residue theorem states that if γ is a semicircle then the integral converges to half this value as $r \to 0$ (see Appendix 2).

Piecing all this together we have

$$0 = \int_{\gamma} \frac{e^{iz}}{z}\, dz = 2i \int_r^R \frac{\sin t}{t}\, dt + \int_{\gamma_2} \frac{e^{iz}}{z}\, dz + \int_{\gamma_4} \frac{e^{iz}}{z}\, dz$$

which on letting $R \to \infty$, $r \to 0$ gives

$$0 = 2i \int_0^{\infty} \frac{\sin t}{t}\, dt - \pi i,$$

and hence

$$\int_0^{\infty} \frac{\sin t}{t}\, dt = \frac{\pi}{2}.$$

Example Consider the integral

$$\int_{-\infty}^{\infty} \left(\frac{\sin x}{x}\right)^2 dx.$$

To evaluate this integral we observe that

$$\int_{-\infty}^{\infty} \left(\frac{\sin x}{x}\right)^2 dx = \frac{1}{2}\int_{-\infty}^{\infty} \frac{1-\cos 2x}{x^2}\,dx = \frac{1}{2}\,\mathrm{Re}\int_{-\infty}^{\infty} \frac{1-e^{2ix}}{x^2}\,dx,$$

and work with

$$\frac{1}{2}\int_{\gamma} \frac{1-e^{2iz}}{z^2}\,dz$$

where $\gamma = \gamma_1 + \gamma_2 + \gamma_3 + \gamma_4$ as in Figure 6.5.

The Laurent expansion of the integrand at $z = 0$ is

$$\frac{1}{2}\frac{1-e^{2iz}}{z^2} = \frac{1}{2}\frac{1-(1+2iz+\cdots)}{z^2} = -\frac{i}{z} + \cdots,$$

which shows that $z = 0$ is a simple pole with residue $-i$. Therefore, the half residue theorem applies and, continuing as in the first example, we obtain

$$\int_{-\infty}^{\infty} \left(\frac{\sin x}{x}\right)^2 dx = \pi.$$

Examples

Evaluate the following integrals.

1. $\displaystyle\int_{-\infty}^{\infty} \frac{dx}{x^2+4}$ $(\pi/2)$
2. $\displaystyle\int_{-\infty}^{\infty} \frac{dx}{(x^2+1)(x^2+4)}$ $(\pi/6)$
3. $\displaystyle\int_{-\infty}^{\infty} \frac{dx}{(x^2+1)^2}$ $(\pi/2)$
4. $\displaystyle\int_{-\infty}^{\infty} \frac{dx}{x^2+x+1}$ $(2\pi/\sqrt{3})$
5. $\displaystyle\int_{-\infty}^{\infty} \frac{\cos x\,dx}{x^2+2x+2}$ $(\pi e^{-1}\cos 1)$

6. $\displaystyle\int_{-\infty}^{\infty} \frac{x^2+1}{x^4+1}\,dx \quad (\pi\sqrt{2})$

7. $\displaystyle\int_{0}^{\infty} \frac{dx}{x^5+1} \quad ((\pi/5)\sin(\pi/5))$

8. $\displaystyle\int_{-\infty}^{\infty} \left(\frac{\sin x}{x}\right)^3 dx \quad (3\pi/4)$

Chapter 7

Summation of series

7.1 Residues of $\cot z$

Elementary theory of sequences and series only allows very few series to be summed exactly. In most cases, one has to be content with knowing that a series converges without knowing what the sum is. It is however possible to sum a wide class of series by exploiting properties of the complex cotangent function $\cot z$.

The singularities of $\cot z = \cos z / \sin z$ occur at the zeros of $\sin z$ which are at $z = n\pi$ for integral n (see Section 2.9). The residues at these singularities can be obtained by differentiating the denominator rule, and are

$$\operatorname*{Res}_{z=n\pi} \cot z = \operatorname*{Res}_{z=n\pi} \frac{\cos z}{\sin z} = \left[\frac{\cos z}{\cos z}\right]_{z=n\pi} = 1.$$

7.2 Laurent expansion of $\cot z$

We can either divide the Maclaurin expansions of $\cos z$, $\sin z$ (as we did in Section 3.6 for $\tan z$) or use the expansion of $\tan z$ to obtain

$$\begin{aligned}\cot z &= \frac{1}{\tan z} = \frac{1}{z + z^3/3 + 2z^5/15 + \cdots} \\ &= \frac{1}{z}\left(1 - \left(\frac{z^2}{3} + \frac{2}{15}z^4 + \cdots\right) + \left(\frac{z^2}{3} + \frac{2}{15}z^4 + \cdots\right)^2 - \cdots\right) \\ &= \frac{1}{z} - \frac{z}{3} - \frac{z^3}{45} - \cdots .\end{aligned}$$

7.3 The method

We demonstrate the method by summing the series

$$\sum_{1}^{\infty} \frac{1}{n^2}.$$

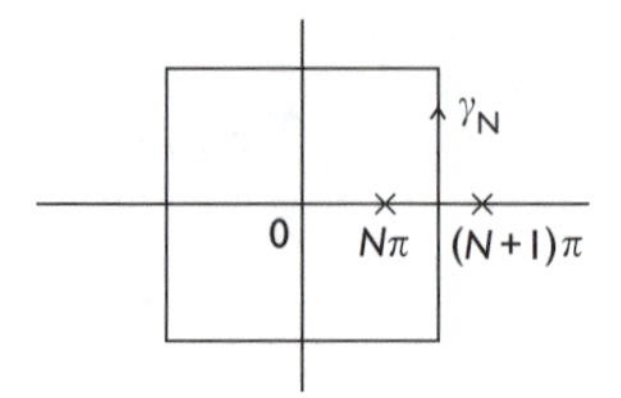

Figure 7.1

We let γ_N be the square centre 0 with half side $(N + 1/2)\pi$ (Figure 7.1), and consider the integral

$$\int_{\gamma_N} \frac{\cot z}{z^2}\, dz.$$

The integrand has singularities at $z = n\pi$ where the residues are

$$\operatorname*{Res}_{z=n\pi} \frac{\cot z}{z^2} = \frac{1}{n^2\pi^2}$$

for $n \neq 0$. At $z = 0$ the Laurent expansion is

$$\frac{\cot z}{z^2} = \frac{1}{z^2}\left(\frac{1}{z} - \frac{z}{3} - \frac{z^3}{45} - \cdots\right) = \frac{1}{z^3} - \frac{1}{3z} - \frac{z}{45} - \cdots$$

showing that there is a triple pole at $z = 0$ with residue $-1/3$.

Therefore by the residue theorem (see Section 4.7) we have

$$\int_{\gamma_N} \frac{\cot z}{z^2}\, dz = 2\pi i \left(2\sum_{1}^{N} \frac{1}{n^2\pi^2} - \frac{1}{3}\right).$$

If we can show the integral $\to 0$ as $N \to \infty$, then we get

$$2\sum_{1}^{N} \frac{1}{n^2\pi^2} - \frac{1}{3} \to 0,$$

and hence

$$\sum_{1}^{\infty} \frac{1}{n^2} = \frac{\pi^2}{6}.$$

7.4 Boundedness of $\cot z$

From Section 2.8 we have

$$|\cot z|^2 = \left|\frac{\cos z}{\sin z}\right|^2 = \frac{\cos^2 x + \sinh^2 y}{\sin^2 x + \sinh^2 y},$$

where $z = x + iy$.

For y fixed we have

$$\frac{\cos^2 x + \sinh^2 y}{\sin^2 x + \sinh^2 y} \le \frac{1 + \sinh^2 y}{\sinh^2 y} = \frac{\cosh^2 y}{\sinh^2 y} = \coth^2 y,$$

which shows that $|\cot z| \le \coth \pi/2 = 1.090331411\ldots$ for all $z \in$ the horizontal sides of γ_N for all $N \ge 1$.

For x fixed $= (N + 1/2)\pi$ we have

$$\frac{\cos^2 x + \sinh^2 y}{\sin^2 x + \sinh^2 y} = \frac{\sinh^2 y}{1 + \sinh^2 y} \le 1,$$

which shows that $|\cot z| \le 1$ for all $z \in$ the vertical sides of γ_N for all $N \ge 1$.

Hence we have $|\cot z| \le M = \coth \pi/2$ for all $z \in \gamma_N$ for all $N \ge 1$.

We can now show

$$\int_{\gamma_N} \frac{\cot z}{z^2}\, dz \to 0$$

as $N \to \infty$. For $z \in \gamma_N$ we have $|z| \ge (N + 1/2)\pi > N\pi$, and the length of γ_N is $8(N + 1/2)\pi \le 9N\pi$ for $N \ge 4$. Therefore by the estimate lemma (4.4) we have

$$\left|\int_{\gamma_N} \frac{\cot z}{z^2}\, dz\right| \le \frac{9MN\pi}{N^2\pi^2} = \frac{9M}{N\pi} \to 0$$

as $N \to \infty$ as required.

7.5 Use of $\operatorname{cosec} z$

Having shown that

$$1 + \frac{1}{2^2} + \frac{1}{3^2} + \frac{1}{4^2} + \cdots = \frac{\pi^2}{6},$$

we can obtain the sum of

$$1 - \frac{1}{2^2} + \frac{1}{3^2} - \frac{1}{4^2} + \cdots = S$$

by observing that

$$\begin{aligned}
S &= \left(1 + \frac{1}{2^2} + \frac{1}{3^2} + \frac{1}{4^2} + \cdots\right) - 2\left(\frac{1}{2^2} + \frac{1}{4^2} + \cdots\right) \\
&= \left(1 + \frac{1}{2^2} + \frac{1}{3^2} + \frac{1}{4^2} + \cdots\right) - \frac{1}{2}\left(1 + \frac{1}{2^2} + \cdots\right) \\
&= \frac{1}{2}\left(1 + \frac{1}{2^2} + \frac{1}{3^2} + \frac{1}{4^2} + \cdots\right) \\
&= \frac{\pi^2}{12}.
\end{aligned}$$

An alternative way to sum this second series is to use $\operatorname{cosec} z$. The singularities of $\operatorname{cosec} z = 1/\sin z$ are at $z = n\pi$ where the residues are

$$\operatorname*{Res}_{z=n\pi} \operatorname{cosec} z = \operatorname*{Res}_{z=n\pi} \frac{1}{\sin z} = \left[\frac{1}{\cos z}\right]_{z=n\pi} = (-1)^n.$$

The Laurent expansion of $\operatorname{cosec} z$ at $z = 0$ is

$$\begin{aligned}
\operatorname{cosec} z &= \frac{1}{\sin z} = \frac{1}{z - z^3/3! + z^5/5! - \cdots} \\
&= \frac{1}{z}\left(1 + \left(\frac{z^2}{6} - \frac{z^4}{120} + \cdots\right) + \left(\frac{z^2}{6} - \frac{z^4}{120} + \cdots\right)^2 + \cdots\right) \\
&= \frac{1}{z} + \frac{z}{6} + \frac{7}{360}z^3 + \cdots.
\end{aligned}$$

And $\operatorname{cosec} z$ is bounded on γ_N as in Section 7.3 since

$$|\operatorname{cosec} z|^2 = \frac{1}{|\sin z|^2} = \frac{1}{\sin^2 x + \sinh^2 y} \le \frac{1}{\sinh^2 y} < \frac{1}{y^2} \le \frac{4}{\pi^2} < 1$$

for $|y| \ge \pi/2$, and for $x = (N + 1/2)\pi$

$$|\operatorname{cosec} z|^2 = \frac{1}{\sin^2 x + \sinh^2 y} = \frac{1}{1 + \sinh^2 y} \le 1.$$

It follows (see Section 7.4) that

$$\left|\int_{\gamma_N} \frac{\operatorname{cosec} z}{z^2}\, dz\right| \le \frac{9N\pi}{N^2\pi^2} = \frac{9}{N\pi} \to 0$$

as $N \to \infty$. But

$$\int_{\gamma_N} \frac{\operatorname{cosec} z}{z^2}\, dz = 2\pi i\left(2\sum_{1}^{\infty} \frac{(-1)^n}{\pi^2 n^2} + \frac{1}{6}\right).$$

Hence we obtain

$$\sum_{1}^{\infty} \frac{(-1)^n}{n^2} = -\frac{\pi^2}{12},$$

equivalently,

$$1 - \frac{1}{2^2} + \frac{1}{3^2} - \frac{1}{4^2} + \cdots = \frac{\pi^2}{12}$$

as before.

7.6 Use of $\tan z$

We can sum the series

$$1 + \frac{1}{3^2} + \frac{1}{5^2} + \frac{1}{7^2} + \cdots$$

given that

$$1 + \frac{1}{2^2} + \frac{1}{3^2} + \frac{1}{4^2} + \cdots = \frac{\pi^2}{6},$$

by observing that

$$\begin{aligned} 1 + \frac{1}{3^2} + \frac{1}{5^2} + \frac{1}{7^2} + \cdots &= \left(1 + \frac{1}{2^2} + \frac{1}{3^2} + \frac{1}{4^2} + \cdots\right) \\ &\quad - \left(\frac{1}{2^2} + \frac{1}{4^2} + \cdots\right) \\ &= \frac{\pi^2}{6} - \frac{1}{4}\frac{\pi^2}{6} \\ &= \frac{\pi^2}{8}. \end{aligned}$$

Alternatively, we can use the integral

$$\int_{\gamma_n} \frac{\tan z}{z^2}\, dz,$$

where γ_N is the square centre at 0 with half side $N\pi$ (Figure 7.2).

The singularities of $\tan z$ are at $z = (n + 1/2)\pi$ with residues -1. Therefore the residue of $\tan z/z^2$ at $z = (n + 1/2)\pi$ is

$$\operatorname*{Res}_{z=(n+1/2)\pi} \frac{\tan z}{z^2} = -\frac{1}{(n + 1/2)^2\pi^2} = -\frac{4}{(2n + 1)^2\pi^2}.$$

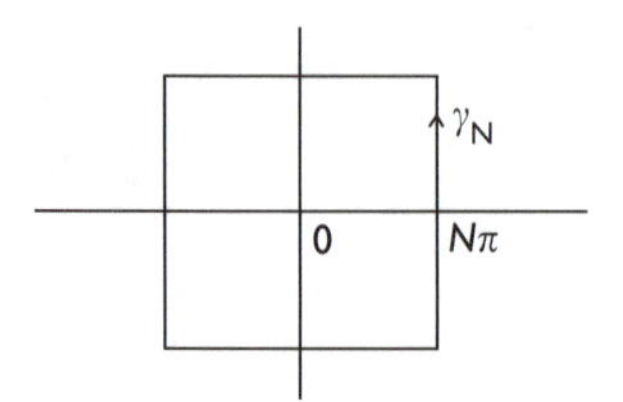

Figure 7.2

There is also a singularity of $\tan z/z^2$ at $z = 0$ where the Laurent expansion is (see Section 3.6)

$$\frac{\tan z}{z^2} = \frac{1}{z^2}\left(z + \frac{z^3}{3} + \frac{2z^5}{15} + \cdots\right) = \frac{1}{z} + \frac{z}{3} + \frac{2z^3}{15} + \cdots.$$

Therefore the residue at $z = 0$ is 1.

It follows that

$$\int_{\gamma_n} \frac{\tan z}{z^2}\, dz = 2\pi i\left(-\sum_{-N}^{N-1} \frac{4}{(2n+1)^2\pi^2} + 1\right),$$

which gives

$$\sum_{-\infty}^{\infty} \frac{1}{(2n+1)^2} = \frac{\pi^2}{4},$$

equivalently,

$$1 + \frac{1}{3^2} + \frac{1}{5^2} + \frac{1}{7^2} + \cdots = \frac{\pi^2}{8},$$

provided we can show

$$\int_{\gamma_n} \frac{\tan z}{z^2}\, dz \to 0$$

as $N \to \infty$.

For this it is sufficient as previously to show $\tan z$ is bounded on γ_N for all N. Which it is since

$$|\tan z|^2 = \left|\frac{\sin z}{\cos z}\right|^2 = \frac{\sin^2 x + \sinh^2 y}{\cos^2 x + \sinh^2 y} \le \frac{1 + \sinh^2 \pi}{\sinh^2 \pi} = \coth^2 \pi$$

for $|y| \geq \pi$. And for $x = N\pi$ we have

$$|\tan z|^2 = \frac{\sinh^2 y}{1 + \sinh^2 y} \leq 1.$$

Hence $|\tan z| \leq \coth \pi$ for all $z \in$ all γ_N.

7.7 Use of $\cot \pi z$

Consider the series

$$\sum_1^\infty \frac{1}{n^2 + 1}.$$

The integral

$$\int_{\gamma_N} \frac{\cot z}{z^2 + 1}\, dz,$$

where γ_N is the square centre 0 with half side $(N + 1/2)\pi$ will sum the series

$$\sum_1^\infty \frac{1}{\pi^2 n^2 + 1}$$

which is not quite what we want. Instead we use

$$\int_{\gamma_N} \frac{\cot \pi z}{z^2 + 1}\, dz,$$

where γ_N is the square centre 0 with half side $(N + 1/2)$ (Figure 7.3).

The singularities of $\cot \pi z$ occur at $z = n$ with residues

$$\operatorname*{Res}_{z=n} \cot \pi z = \operatorname*{Res}_{z=n} \frac{\cos \pi z}{\sin \pi z} = \left[\frac{\cos \pi z}{\pi \cos \pi z} \right]_{z=n} = \frac{1}{\pi}.$$

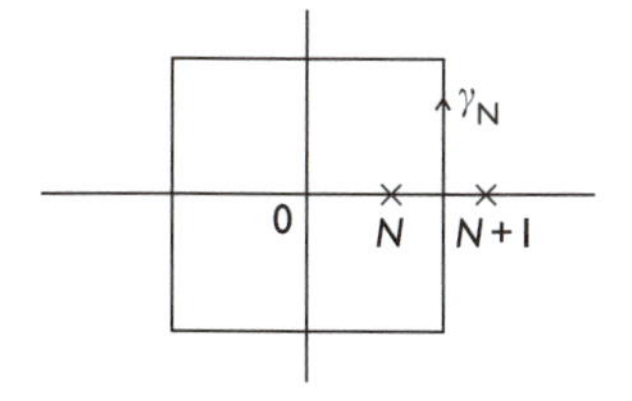

Figure 7.3

Therefore,

$$\operatorname*{Res}_{z=n} \frac{\cot \pi z}{z^2+1} = \frac{1}{\pi} \frac{1}{n^2+1}.$$

The integrand also has singularities at $z = \pm i$ where the residues are

$$\operatorname*{Res}_{z=i} \frac{\cot \pi z}{z^2+1} = \left[\frac{\cot \pi z}{2z} \right]_{z=i} = \frac{\cot \pi i}{2i} = -\frac{\coth \pi}{2} = \operatorname*{Res}_{z=-i} \frac{\cot \pi z}{z^2+1}.$$

Therefore,

$$\int_{\gamma_N} \frac{\cot \pi z}{z^2+1}\, dz = 2\pi i \left(\frac{1}{\pi} \sum_{-N}^{N} \frac{1}{n^2+1} - \coth \pi \right),$$

which gives

$$\sum_{-\infty}^{\infty} \frac{1}{n^2+1} = \pi \coth \pi,$$

provided the integral $\to 0$ as $N \to \infty$. Which it does since $|\coth \pi z| \leq \coth \pi/2$ on γ_N, and the length of γ_N is $8(N+1/2)$, therefore

$$\left| \int_{\gamma_N} \frac{\cot \pi z}{z^2+1}\, dz \right| \leq \frac{8(N+1/2)\coth \pi/2}{(N+1/2)^2 - 1} \to 0$$

as $N \to \infty$.

7.8 Use of $\sec z$

It might be thought that the integral

$$\int_{\gamma_N} \frac{\sec z}{z^2}\, dz,$$

where γ_N is the square centre 0 with half side $N\pi$ (see Figure 7.2) will sum the series

$$1 - \frac{1}{3^2} + \frac{1}{5^2} - \frac{1}{7^2} + \cdots$$

However it turns out that it doesn't. The problem is that it sums the series

$$\sum_{-\infty}^{\infty} \frac{(-1)^n}{(2n+1)^2} = 0$$

which is true but not very helpful.

In fact, no closed form is known for the sum of the series

$$1-\frac{1}{3^2}+\frac{1}{5^2}-\frac{1}{7^2}+\cdots$$

Examples

1. Find the sum of the series

$$1+\frac{1}{2^4}+\frac{1}{3^4}+\frac{1}{4^4}+\cdots$$

by integrating $\cot z/z^4$ round a large square contour.

2. Use your answer to Question 1 to find the sums of the following series.

$$1-\frac{1}{2^4}+\frac{1}{3^4}-\frac{1}{4^4}+\cdots$$

$$1+\frac{1}{3^4}+\frac{1}{5^4}+\frac{1}{7^4}+\cdots.$$

3. Find the sum of

$$1-\frac{1}{2^4}+\frac{1}{3^4}-\frac{1}{4^4}+\cdots$$

by integrating $\operatorname{cosec} z/z^4$ round a large square contour. Compare your answer with the answer you got in Question 2.

4. Find the sum of

$$1+\frac{1}{3^4}+\frac{1}{5^4}+\frac{1}{7^4}+\cdots$$

by integrating $\tan z/z^4$ round a large square contour. Compare with the answer you got in Question 2.

5. Find the sum of

$$1+\frac{1}{3^4}+\frac{1}{5^4}+\frac{1}{7^4}+\cdots$$

by integrating $\cot \pi z/(2z+1)^4$ round a large square contour. Compare with Questions 2 and 4.

6. Find the sum of

$$\sum_{1}^{\infty}\frac{1}{n(n+1)}$$

by writing the nth term as

$$\frac{1}{n(n+1)} = \frac{1}{n} - \frac{1}{n+1}.$$

7. Find the sum of

$$\sum_{1}^{\infty} \frac{1}{n(n+1)}$$

by integrating $\cot \pi z / z(z+1)$ round a large square contour. Compare with Question 6.

Chapter 8

Fundamental theorem of algebra

8.1 Zeros

We call the point c a *zero* of the function $f(z)$ if $f(c) = 0$. For example, the zeros of $\sin z$ are at $z = n\pi$ for $n = 0, \pm 1, \pm 2, \ldots$.

We define the *order* or *multiplicity* of a zero as follows. Suppose $f(z)$ has Taylor expansion

$$f(z) = \sum_{0}^{\infty} a_n (z - c)^n = \sum_{0}^{\infty} \frac{f^{(n)}(c)}{n!}(z - c)^n$$

at $z = c$. We say c is a zero of order n if $a_0 = a_1 = \cdots = a_{n-1} = 0$, but $a_n \neq 0$. Equivalently, if $f(c) = f'(c) = \cdots = f^{(n-1)}(c) = 0$, but $f^{(n)}(c) \neq 0$. A zero of order 1 is called a *simple* zero, a zero of order 2 is called a *double* zero, etc. For example, the zeros of $f(z) = \sin z$ are all simple since $f'(z) = \cos z = \pm 1$ at $z = n\pi$. However, for example, $g(z) = z \sin z$ has a double zero at $z = 0$ since the Maclaurin expansion is

$$z \sin z = z\left(z - \frac{z^3}{3!} + \cdots\right) = z^2 - \frac{z^4}{3!} + \cdots$$

Theorem 1 (Fundamental theorem of algebra) Every polynomial of degree n with complex coefficients has n zeros in the complex plane taking account of multiplicity.

Case $n = 2$ Every quadratic polynomial $p(z)$ with complex coefficients has 2 roots, possibly coincident. The case of coincident roots is when $p(z)$ is a perfect square taking the form

$$p(z) = A(z - B)^2,$$

therefore $p(z)$ has a double zero at $z = B$.

8.2 Argument principle

We can count the number of zeros a function has inside a closed contour by means of the following theorem.

Theorem 2 (Argument principle) If $f(z)$ is differentiable inside and on the closed contour γ, and if $f(z) \neq 0$ anywhere on γ, then the number N of zeros of $f(z)$ inside γ is given by the formula

$$N = \frac{1}{2\pi i} \int_{\gamma} \frac{f'(z)}{f(z)}\, dz.$$

Geometrical interpretation Observe that

$$\frac{1}{2\pi i} \int_{\gamma} \frac{f'(z)}{f(z)}\, dz = \frac{1}{2\pi i} \left[\log f(z)\right]_{\gamma}$$

since $\log f(z)$ is a primitive of $f'(z)/f(z)$. But (see Section 2.12)

$$\log f(z) = \log |f(z)| + i \arg f(z),$$

and $\log |f(z)|$ is single valued. Therefore,

$$\frac{1}{2\pi i} \left[\log f(z)\right]_{\gamma} = \frac{1}{2\pi} \left[\arg f(z)\right]_{\gamma}.$$

So Theorem 2 says that the number of zeros of $f(z)$ inside γ is equal to the number of times $f(z)$ circulates the origin as z goes round γ.

Example Suppose $f(z) = z^2 - 1$.

Case 1 γ_1 = circle centre 0, radius 1/2.

We can parametrise γ_1 as $z = e^{it}/2$ $(0 \leq t \leq 2\pi)$. Therefore, the image contour $f(\gamma_1)$ parametrises as $w = f(z) = z^2 - 1 = e^{2it}/4 - 1$ which is the circle centre -1, radius 1/4 described twice. Observe that $f(\gamma_1)$ does not circulate the origin at all, corresponding to the fact that there are no zeros of $f(z)$ inside γ_1 (Figure 8.1).

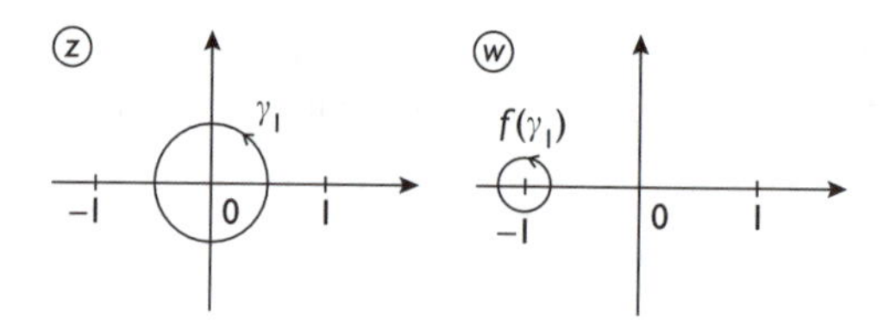

Figure 8.1

Case 2 $\gamma_2 =$ circle centre 0, radius 2.

We can parametrise γ_2 as $z = 2e^{it}(0 \leq t \leq 2\pi)$. Therefore, the image contour $f(\gamma_2)$ parametrises as $w = f(z) = z^2 - 1 = 4e^{2it} - 1$ which is the circle centre -1, radius 4 described twice. Hence in this case the image contour $f(\gamma_2)$ circulates the origin twice, reflecting the fact that $f(z)$ has 2 zeros inside γ_2 (Figure 8.2).

Case 3 $\gamma_3 =$ circle centre 1, radius 1.

On γ_3 we have $z = 1 + e^{it}(0 \leq t \leq 2\pi)$. Therefore on $f(\gamma_3)$ we have

$$w = f(z) = (1 + e^{2it})^2 - 1 = e^{it}(e^{-it/2} + e^{it/2})^2 - 1 = 4e^{it}\cos^2 t/2 - 1$$

which shows $w + 1 = re^{i\theta}$ where $r = 4\cos^2 t/2, \theta = t$. Hence in this case the image contour is the cardioid illustrated in Figure 8.3 which circulates the origin once, in agreement with the fact that $z^2 - 1$ has one zero inside γ at $z = 1$.

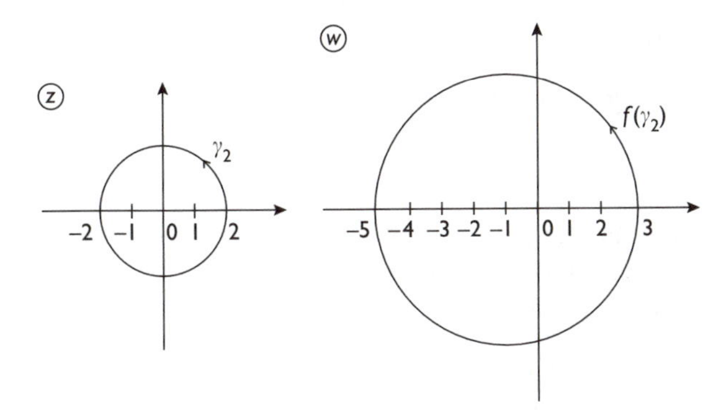

Figure 8.2

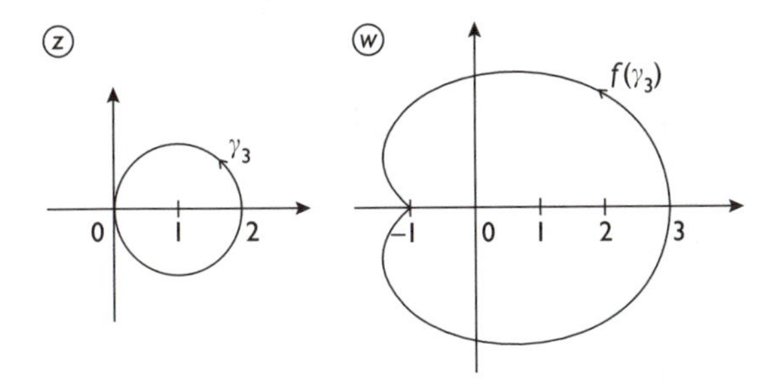

Figure 8.3

Proof of the argument principle The singularities of $f'(z)/f(z)$ occur at the zeros of $f(z)$. If $f(z)$ has a zero of order n at $z = c$, then the Taylor expansion is

$$f(z) = a_n(z-c)^n + a_{n+1}(z-c)^{n+1} + \cdots,$$

where $a_n \neq 0$. Therefore,

$$\begin{aligned}\frac{f'(z)}{f(z)} &= \frac{na_n(z-c)^{n-1} + (n+1)a_{n+1}(z-c)^n + \cdots}{a_n(z-c)^n + a_{n+1}(z-c)^{n+1} + \cdots} \\ &= \frac{1}{z-c}\frac{na_n + (n+1)a_{n+1}(z-c) + \cdots}{a_n + a_{n+1}(z-c) + \cdots}\end{aligned}$$

has a simple pole at $z = c$ with residue n by the cover up rule (see Section 4.8). The result follows.

8.3 Rouché's theorem

The following theorem due to Rouché (1862) enables us to say something about the distribution of the zeros of a given function by comparing it with another function whose zeros are known.

Theorem 3 (Rouché's theorem) If $f(z)$, $g(z)$ are differentiable inside and on the closed contour γ, and if $|f(z)| > |g(z)|$ for all $z \in \gamma$, then $f(z)$, $f(z) + g(z)$ have the same number of zeros inside γ.

Proof Informally, we can add any 'smaller' function $g(z)$ to $f(z)$ without changing the number of zeros inside the contour.

By the argument principle it will be sufficient to prove that

$$\int_\gamma \frac{f'(z) + g'(z)}{f(z) + g(z)}\,dz = \int_\gamma \frac{f'(z)}{f(z)}\,dz.$$

Observe that

$$\begin{aligned}\frac{f'(z) + g'(z)}{f(z) + g(z)} - \frac{f'(z)}{f(z)} &= \frac{d}{dz}\log(f(z) + g(z)) - \frac{d}{dz}\log f(z) \\ &= \frac{d}{dz}\log\left(\frac{f(z) + g(z)}{f(z)}\right) \\ &= \frac{d}{dz}\log\left(1 + \frac{g(z)}{f(z)}\right) \\ &= \frac{d}{dz}\log h(z) \\ &= \frac{h'(z)}{h(z)}\end{aligned}$$

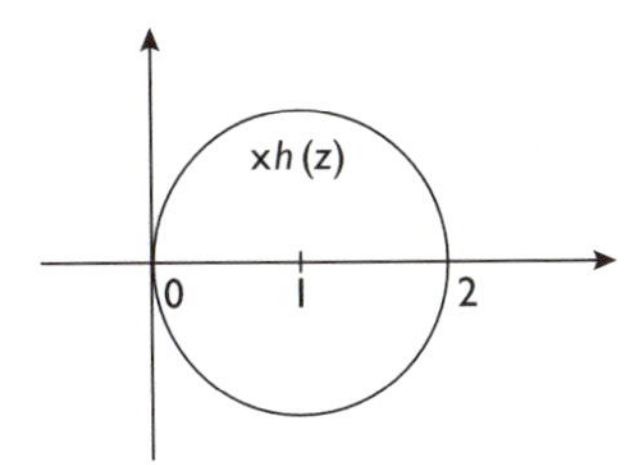

Figure 8.4

where

$$h(z) = 1 + \frac{g(z)}{f(z)}.$$

The condition $|f(z)| > |g(z)|$ implies that $h(z)$ must satisfy the inequality $|h(z) - 1| < 1$ for all $z \in \gamma$. It follows that $h(z)$ cannot circulate the origin as z goes round γ (Figure 8.4). Therefore by the argument principle we must have

$$\int_\gamma \frac{h'(z)}{h(z)}\, dz = 0$$

as required.

Application We can use Rouché's theorem to show, for example, that the zeros of the polynomial $p(z) = z^3 + z^2 + 3$ all lie in the annulus $1 < |z| < 2$. If we take $f(z) = z^3$, $g(z) = z^2 + 3$, then for $|z| = 2$ we have

$$|g(z)| = |z^2 + 3| \le |z^2| + 3 = 4 + 3 = 7 < 8 = |z^3| = |f(z)|.$$

Therefore by Rouché's theorem $p(z) = f(z) + g(z)$ and $f(z)$ have the same number of zeros inside $|z| = 2$. But $f(z) = z^3$ has 3 zeros inside $|z| = 2$ in the form of a triple zero at $z = 0$. Hence also $p(z)$ has 3 zeros inside $|z| = 2$.

If instead we take $f(z) \equiv 3$, $g(z) = z^3 + z^2$, then for $|z| = 1$ we have

$$|g(z)| = |z^3 + z^2| \le |z^3| + |z^2| = 1 + 1 = 2 < 3 = |f(z)|.$$

Therefore by Rouché's theorem $p(z) = f(z) + g(z)$ and $f(z) \equiv 3$ have the same number of zeros inside $|z| = 1$. But $f(z)$ has no zeros inside $|z| = 1$. Hence neither has $p(z)$ (Figure 8.5).

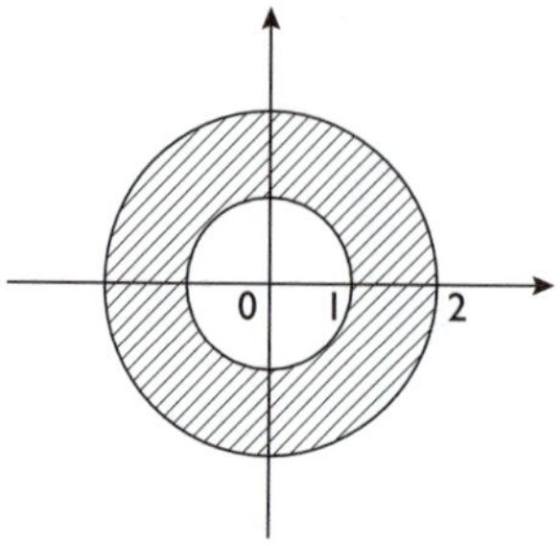

Figure 8.5

8.4 Proof of the fundamental theorem of algebra

Suppose that

$$p(z) = a_n z^n + a_{n-1} z^{n-1} + \cdots + a_1 z + a_0,$$

where $a_n \neq 0$, is a polynomial of degree n. Let

$$f(z) = a_n z^n,$$
$$g(z) = a_{n-1} z^{n-1} + \cdots + a_1 z + a_0,$$

and let γ_R be the contour $|z| = R$. Then on γ_R we have

$$\begin{aligned}\left|\frac{g(z)}{f(z)}\right| &= \left|\frac{a_{n-1} z^{n-1} + \cdots + a_1 z + a_0}{a_n z^n}\right| \\ &\leq \frac{|a_{n-1} z^{n-1}| + \cdots + |a_1 z| + |a_0|}{|a_n z^n|} \\ &= \frac{|a_{n-1}| R^{n-1} + \cdots + |a_1| R + |a_0|}{|a_n| R^n} \\ &\to 0\end{aligned}$$

as $R \to \infty$. Therefore, we can choose R such that $|f(z)| > |g(z)|$ for all $z \in \gamma_R$.

It follows by Rouché's theorem that $p(z) = f(z) + g(z)$ and $f(z)$ have the same number of zeros inside γ_R for this R. But $f(z) = a_n z^n$ has n zeros inside γ_R, all at $z = 0$. Hence also $p(z)$ has n zeros inside γ_R, as required.

Examples

1. Prove that all the zeros of the polynomial $z^3 + 9z^2 + 9z + 9$ lie inside the circle $|z| = 10$.

2. Prove that exactly two zeros of the polynomial $z^3 + 9z^2 + 9z + 9$ lie inside the circle $|z| = 2$.
3. Prove that none of the zeros of the polynomial $z^3 + 9z^2 + 9z + 9$ lie inside the circle $|z| = 1/2$.
4. Prove that all the zeros of the polynomial $z^3 + 6z + 8$ lie between the two circles $|z| = 1$, $|z| = 3$.
5. Prove that the polynomial $z^4 + z + 1$ has one zero in each quadrant.

Solutions to examples

1 Complex numbers

1. (i) $6 + 10i$. (ii) $-4 - 4i$. (iii) $-16 + 22i$. (iv) $(13/37) + (4/37)i$. (v) $\pm(2 + i)$. (vi) $\log\sqrt{2} + i\pi/4$ (PV).

2. $$\sqrt{1+i} = \pm\sqrt{\frac{\sqrt{2}+1}{2}} \pm i\sqrt{\frac{\sqrt{2}-1}{2}} = \pm\sqrt[4]{2}e^{i\pi/8}.$$

   Since $\arg\sqrt{1+i} = \pi/8$ (PV), we have

   $$\tan\pi/8 = \sqrt{\frac{\sqrt{2}-1}{\sqrt{2}+1}} = \sqrt{2} - 1 = 0.4142135624\ldots$$

3. If $C = \cos\theta$, $S = \sin\theta$, then

   $$\cos 3\theta = C^3 - 3CS^2 = C^3 - 3C(1 - C^2) = 4C^3 - 3C,$$
   $$\sin 3\theta = 3C^2S - S^3 = 3(1 - S^2)S - S^3 = 3S - 4S^3.$$

4. If $C = \cos 30°$, then

   $$0 = \cos 90° = 4C^3 - 3C = C(4C^2 - 3).$$

   Therefore $C = 0$ or $\pm\sqrt{3}/2$. What do the other values of C represent?

5. $(e^{i\theta} + e^{-i\theta})^3 = e^{3i\theta} + 3e^{i\theta} + 3e^{-i\theta} + e^{-3i\theta} = 2\cos 3\theta + 6\cos\theta$.

6. $$\int_0^{\pi/2} \cos^3\theta \, d\theta = \int_0^{\pi/2} \sin^3\theta \, d\theta = 2/3.$$

7. $(e^{2\pi} - 1)/10$.

2 Complex functions

8. To find the maximum of $|\sin z|$ on the disc $|z| \leq 1$ we use the infinite form of the triangle inequality which states that

$$\left|\sum_{1}^{\infty} z_n\right| \leq \sum_{1}^{\infty} |z_n|$$

for any sequence of complex numbers $(z_n)_{n\geq 1}$.

In particular,

$$\begin{aligned}
|\sin z| &= \left|z - \frac{z^3}{3!} + \frac{z^5}{5!} - \frac{z^7}{7!} + \cdots\right| \\
&\leq |z| + \frac{|z|^3}{3!} + \frac{|z|^5}{5!} + \frac{|z|^7}{7!} + \cdots \\
&\leq 1 + \frac{1}{3!} + \frac{1}{5!} + \frac{1}{7!} + \cdots \\
&= \sinh 1
\end{aligned}$$

for $|z| \leq 1$.

Also for $z = i$ we have $|\sin i| = |i \sinh 1| = \sinh 1$.

Therefore, $|\sin z|$ is maximum on $|z| \leq 1$ at $z = i$ with maximum value equal to $\sinh 1$.

9. If $z = x + iy$ then

$$\left|\frac{z+1}{z+4}\right| = 2$$

is equivalent to

$$\begin{aligned}
|x + iy + 1|^2 &= 4|x + iy + 4|^2 \\
(x+1)^2 + y^2 &= 4((x+4)^2 + y^2) \\
x^2 + 2x + 1 + y^2 &= 4(x^2 + 8x + 16 + y^2) \\
0 &= 3x^2 + 3y^2 + 30x + 63 \\
0 &= x^2 + y^2 + 10x + 21 \\
4 &= x^2 + 10x + 25 + y^2 = (x+5)^2 + y^2 \\
4 &= |x + 5 + iy|^2 \\
2 &= |z + 5|
\end{aligned}$$

which is the equation of a circle centre -5, radius 2.

3 Derivatives

1. If $z = re^{i\theta} = x + iy$, then

$$\log z = \log r + i\theta = \tfrac{1}{2}\log(x^2 + y^2) + i\tan^{-1}\frac{y}{x}.$$

$$u(x, y) = \tfrac{1}{2}\log(x^2 + y^2), \quad v(x, y) = \tan^{-1}\frac{y}{x}.$$

$$\frac{\partial u}{\partial x} = \frac{\partial v}{\partial y} = \frac{x}{x^2 + y^2}, \quad \frac{\partial u}{\partial y} = -\frac{\partial v}{\partial x} = \frac{y}{x^2 + y^2}.$$

$$f'(z) = \frac{\partial u}{\partial x} + i\frac{\partial v}{\partial x} = \frac{x - iy}{x^2 + y^2} = \frac{1}{x + iy} = \frac{1}{z}.$$

2. $|z|^2 = x^2 + y^2$. Therefore $u(x, y) = x^2 + y^2$, $v(x, y) = 0$. The Cauchy–Riemann equations hold only at $x = y = 0$. The derivative at $z = 0$ is 0.

3. If $f(z) = \bar{z}(|z|^2 - 2) = (x - iy)(x^2 + y^2 - 2)$, then

$$u(x, y) = x(x^2 + y^2 - 2), \quad v(x, y) = -y(x^2 + y^2 - 2).$$

$$\frac{\partial u}{\partial x} = 3x^2 + y^2 - 2, \quad \frac{\partial u}{\partial y} = 2xy, \quad \frac{\partial v}{\partial x} = -2xy, \quad \frac{\partial v}{\partial y} = -x^2 - 3y^2 + 2.$$

$$\frac{\partial u}{\partial y} = -\frac{\partial v}{\partial x} \quad \text{for all } x, y.$$

$$\frac{\partial u}{\partial x} = \frac{\partial v}{\partial y} \text{ only when}$$

$$3x^2 + y^2 - 2 = -x^2 - 3y^2 + 2,$$

which simplifies to

$$x^2 + y^2 = 1.$$

For $|z| = 1$ we have

$$f'(z) = \frac{\partial u}{\partial x} + i\frac{\partial v}{\partial x} = 3x^2 + y^2 - 2 - 2ixy = x^2 - y^2 - 2ixy = (x - iy)^2 = \bar{z}^2.$$

4. If $f(z)$ is real valued, then $v(x, y) = 0$. Therefore,

$$\frac{\partial u}{\partial x} = \frac{\partial v}{\partial y} = 0, \quad \frac{\partial u}{\partial y} = -\frac{\partial v}{\partial x} = 0.$$

Therefore $u(x, y) = \text{constant}$.

5. $e^z \sin z = z + z^2 + z^3/3 - z^5/30 + \cdots$.

6. (i) Put $t = z - 2$. Then $z = t + 2$. Therefore,

$$\frac{1}{z} = \frac{1}{t+2} = \frac{1}{2}\frac{1}{1+t/2} = \frac{1}{2}\left(1 - \frac{t}{2} + \frac{t^2}{4} - \frac{t^3}{8} + \cdots\right)$$
$$= \frac{1}{2} - \frac{1}{4}(z-2) + \frac{1}{8}(z-2)^2 - \cdots .$$

The range of validity of the geometric series is $|t/2| < 1$. Therefore the range of validity of the Taylor series is $|z - 2| < 2$.

(ii) Put $t = z - i$. Then $z = t + i$. Therefore,

$$e^z = e^{t+i} = e^i e^t = e^i\left(1 + t + \frac{t^2}{2!} + \cdots\right)$$
$$= e^i + e^i(z-i) + \frac{e^i}{2!}(z-i)^2 + \cdots .$$

Range of validity is all t, therefore all z.

(iii) Put $t = z - 1$. Then $z = t + 1$. Therefore,

$$\log z = \log(t+1) = t - \frac{t^2}{2} + \frac{t^3}{3} - \cdots$$
$$= (z-1) - \frac{1}{2}(z-1)^2 + \frac{1}{3}(z-1)^3 - \cdots .$$

Valid for $|t| < 1$, therefore $|z - 1| < 1$.

7. (i) $\dfrac{e^z}{z^{10}} = \dfrac{1}{z^{10}}\left(1 + z + \dfrac{z^2}{2!} + \cdots\right) = \dfrac{1}{z^{10}} + \dfrac{1}{z^9} + \cdots + \dfrac{1}{9!}\dfrac{1}{z} + \cdots .$

Pole of order 10 with residue 1/9!

(ii) $\dfrac{\sin z}{z^{15}} = \dfrac{1}{z^{15}}\left(z - \dfrac{z^3}{3!} + \cdots\right) = \dfrac{1}{z^{14}} - \dfrac{1}{3!}\dfrac{1}{z^{12}} + \cdots + \dfrac{1}{13!}\dfrac{1}{z^2} - \dfrac{1}{15!} + \cdots .$

Pole of order 14 with residue 0.

(iii) To find the Laurent expansion at $z = 1$ put $t = z - 1$. Then

$$\frac{1}{z^2-1} = \frac{1}{(t+1)^2-1} = \frac{1}{t^2+2t} = \frac{1}{2t}\frac{1}{1+t/2}$$
$$= \frac{1}{2t}\left(1 - \frac{t}{2} + \frac{t^2}{4} - \cdots\right) = \frac{1}{2t} - \frac{1}{4} + \frac{t}{8} - \cdots .$$

Simple pole with residue $1/2$.

To find the Laurent expansion at $z = -1$ put $t = z + 1$. Then

$$\frac{1}{z^2 - 1} = \frac{1}{(t-1)^2 - 1} = \frac{1}{t^2 - 2t} = -\frac{1}{2t}\frac{1}{1 - t/2}$$

$$= -\frac{1}{2t}\left(1 + \frac{t}{2} + \frac{t^2}{4} + \cdots\right) = -\frac{1}{2t} - \frac{1}{4} - \frac{t}{8} - \cdots .$$

Simple pole with residue $-1/2$.

8. Using the geometric series expansion we have

$$\frac{3z+1}{(z+2)(z-3)} = \frac{1}{z+2} + \frac{2}{z-3} = \frac{1}{2}\frac{1}{1+z/2} - \frac{2}{3}\frac{2}{1-z/3}$$

$$= \frac{1}{2}\left(1 - \frac{z}{2} + \frac{z^2}{4} - \cdots\right) - \frac{2}{3}\left(1 + \frac{z}{3} + \frac{z^2}{9} + \cdots\right).$$

The first bracket is valid for $|z/2| < 1$, that is, $|z| < 2$, the second for $|z/3| < 1$, that is, $|z| < 3$. Therefore both are valid for $|z| < 2$.

4 Integrals

1. (i) $\displaystyle\int_\gamma \operatorname{Re} z\, dz = \int_0^{2\pi} (\cos t) i e^{it}\, dt = \pi i.$

   (ii) $\displaystyle\int_\gamma |z|^2\, dz = \int_0^1 (t^2 + t^4)(1 + 2it)\, dt = \frac{8}{15} + \frac{5}{6}i.$

   (iii) γ parametrises as $z = (1+i)t$ $(0 \le t \le 1)$. Therefore $dz = (1+i)\, dt$, and hence

$$\int_\gamma \bar{z}\, dz = \int_0^1 (1-i)t(1+i)\, dt = 2\int_0^1 t\, dt = 1.$$

2. (i) $|e^z| = e^{\operatorname{Re} z} \le e^2$ for $z \in \gamma$. Length of γ is 2π.

   (ii) $|\sin z| \le \sinh 1$, $|z + i| \ge \sqrt{2}$ for $z \in \gamma$. Length of γ is π.

   (iii) $|z - 2| \le \sqrt{10}$, $|z - 3| \ge 2$ for $z \in \gamma$. Length of γ is 8.

3. (i) Singularity at 1. Residue 2.

   (ii) Singularities at $\pm\pi i$. Residues $\mp 1/2\pi i$.

   (iii) Singularities at 2,4. Residues $\mp 1/2$.

4. (i) $4\pi i$. (ii) -1. (iii) $-\pi i$.

5. Residue of $\dfrac{f(z)}{z-a}$ at $z = a$ is $f(a)$.

The integral vanishes for a outside γ by Cauchy's theorem.

5 Evaluation of finite real integrals

1. We have

$$\int_0^{2\pi} \frac{dt}{2+\cos t} = \frac{2}{i}\int_\gamma \frac{dz}{z^2+4z+1} = \frac{2\pi}{\sqrt{3}}$$

by the residue theorem. The integrand has one singularity inside γ at $z = \sqrt{3}-2$, where the residue is $1/2\sqrt{3}$ (differentiate the denominator).

2. We have

$$\int_0^{2\pi} \frac{dt}{3+2\sin t} = \int_\gamma \frac{dz}{z^2+3iz-1} = \frac{2\pi}{\sqrt{5}}.$$

The integrand has one singularity inside γ at $z = (\sqrt{5}-3)i/2$, where the residue is $1/\sqrt{5}i$.

3. We have

$$\int_0^{2\pi} \frac{dt}{4-3\cos^2 t} = 4i\int_\gamma \frac{z\,dz}{3z^4-10z^2+3} = \pi.$$

The integrand has two singularities inside γ at $z = \pm 1/\sqrt{3}$, where the residues are both equal to $-1/16$.

4. We have

$$\begin{aligned}\int_0^{2\pi} \frac{\sin 5t}{\sin t}\,dt &= \int_\gamma \frac{z^5-1/z^5}{z-1/z}\frac{dz}{iz} = \int_\gamma \frac{z^{10}-1}{z^2-1}\frac{dz}{iz^5} \\ &= \int_\gamma \frac{1+z^2+z^4+z^6+z^8}{iz^5}\,dz \\ &= \frac{1}{i}\int_\gamma \left(\frac{1}{z^5}+\frac{1}{z^3}+\frac{1}{z}+z+z^3\right)dz = 2\pi.\end{aligned}$$

5. We have

$$\begin{aligned}\int_0^{2\pi} \cos^6 t\,dt &= \frac{1}{64}\int_\gamma \left(z+\frac{1}{z}\right)^6 \frac{dz}{iz} \\ &= \frac{1}{64}\int_\gamma \left(z^6+6z^4+15z^2+20+\frac{15}{z^2}+\frac{6}{z^4}+\frac{1}{z^6}\right)\frac{dz}{iz} = \frac{5\pi}{8}.\end{aligned}$$

6 Evaluation of infinite real integrals

1. $\dfrac{1}{z^2+4}$ has singularities at $z = \pm 2i$.

   The residue at $z = 2i$ is $1/4i$ (differentiate the denominator).

$$\left|\int_{\gamma_2} \frac{dz}{z^2+4}\right| \leq \frac{\pi R}{R^2-4} \, (R > 2) \to 0 \quad \text{as } R \to \infty.$$

2. $\dfrac{1}{(z^2+1)(z^2+4)}$ has singularities at $z = \pm i, \pm 2i$.

   The residue at $z = i$ is $1/6i$. The residue at $z = 2i$ is $-1/12i$.

$$\left|\int_{\gamma_2} \frac{dz}{(z^2+1)(z^2+4)}\right| \leq \frac{\pi R}{(R^2-1)(R^2-4)} \, (R > 2) \to 0 \quad \text{as } R \to \infty.$$

3. $\dfrac{1}{(z^2+1)^2}$ has a double pole at $z = i$.

   To get the residue we have to compute the Laurent expansion. Put $t = z - i$. Then we have

$$\frac{1}{(z^2+1)^2} = \frac{1}{(t^2+2it)^2} = -\frac{1}{4t^2}\left(1+\frac{t}{2i}\right)^{-2}$$
$$= -\frac{1}{4t^2}\left(1-\frac{t}{i}+\cdots\right) = -\frac{1}{4t^2} + \frac{1}{4it} + \cdots$$

(using the binomial theorem with exponent -2). Therefore the residue is $1/4i$.

$$\left|\int_{\gamma_2} \frac{dz}{(z^2+1)^2}\right| \leq \frac{\pi R}{(R^2-1)^2} \, (R > 1) \to 0 \quad \text{as } R \to \infty.$$

4. $\dfrac{1}{z^2+z+1}$ has singularities at $z = \omega, \omega^2$, where $\omega^3 = 1$.

   The residue at $z = \omega$ is $\dfrac{1}{2\omega+1} = \dfrac{1}{\sqrt{3}i}$.

$$\left|\int_{\gamma_2} \frac{dz}{z^2+z+1}\right| \leq \frac{\pi R}{R^2-R-1} \, (R > 2) \to 0 \quad \text{as } R \to \infty.$$

5. Observe first that

$$\int_{-\infty}^{\infty} \frac{\cos x \, dx}{x^2+2x+2} = \text{Re} \int_{-\infty}^{\infty} \frac{e^{ix} \, dx}{x^2+2x+2}.$$

Second that $|e^{iz}| = |e^{i(x+iy)}| = |e^{ix-y}| = |e^{ix}e^{-y}| = e^{-y} \leq 1$ on γ_2. Therefore

$$\left|\int_{\gamma_2} \frac{e^{iz}dz}{z^2+2z+2}\right| \leq \frac{\pi R}{R^2-2R-2}\,(R>3) \to 0 \quad \text{as } R \to \infty.$$

$\dfrac{e^{iz}}{z^2+2z+2}$ has singularities at $z = -1 \pm i$.

The residue at $z = -1+i$ is $e^{-1-i}/2i$. Therefore,

$$\int_{-\infty}^{\infty} \frac{e^{ix}\,dx}{x^2+2x+2} = \pi e^{-1-i}.$$

Now take real parts.

6. $\dfrac{z^2+1}{z^4+1}$ has singularities at $z = \omega, \omega^3, \omega^5, \omega^7$ where $\omega^8 = 1$.

The residue at $z = \omega$ is

$$\frac{\omega^2+1}{4\omega^3} = \frac{\omega^7+\omega^5}{4} = -\frac{\sqrt{2}i}{4} = -\frac{i}{2\sqrt{2}}.$$

The residue at $z = \omega^3$ is

$$\frac{\omega^6+1}{4\omega^9} = \frac{\omega^6+1}{4\omega} = \frac{\omega^5+\omega^7}{4} = -\frac{\sqrt{2}i}{4} = -\frac{i}{2\sqrt{2}}.$$

$$\left|\int_{\gamma_2} \frac{z^2+1}{z^4+1}\,dz\right| \leq \frac{\pi R(R^2+1)}{R^4-1}\,(R>1) \to 0 \quad \text{as } R \to \infty.$$

7. Use a pizza slice contour with angle $2\pi/5$.

8. Observe that for x real

$$\left(\frac{\sin x}{x}\right)^3 = \frac{3\sin x - \sin 3x}{4x^3} = \operatorname{Im}\frac{3e^{ix}-e^{3ix}}{4x^3},$$

and that for z complex

$$\frac{3e^{iz}-e^{3iz}}{4z^3} = \frac{3(1+iz-z^2/2!+\cdots)-(1+3iz-9z^2/2!+\cdots)}{4z^3}$$

$$= \frac{2+3z^2+\cdots}{4z^3} = \frac{1}{2z^3} + \frac{3}{4z} + \cdots$$

(See N.B. at the end of Appendix 2.)

7 Summation of series

1. $\pi^4/90$.

2. $7\pi^4/720$, $\pi^4/96$.

6. Observe that

$$\sum_1^{\infty}\frac{1}{n(n+1)}=\sum_1^{\infty}\left(\frac{1}{n}-\frac{1}{n+1}\right)$$
$$=\left(1-\frac{1}{2}\right)+\left(\frac{1}{2}-\frac{1}{3}\right)+\cdots+\left(\frac{1}{n}-\frac{1}{n+1}\right)+\cdots=1.$$

7. If γ is the square with centre 0, and half side $N+1/2$, then

$$\left|\int_\gamma \frac{\cot \pi z}{z(z+1)}\,dz\right|\le\frac{8(N+1/2)\coth \pi/2}{(N+1/2)(N+3/2)}=\frac{16\coth \pi/2}{2N+3}\to 0$$

as $N\to\infty$. For $n\neq 0,-1$ we have

$$\operatorname*{Res}_{z=n}\frac{\cot \pi z}{z(z+1)}=\frac{1}{\pi n(n+1)}.$$

At $z=0$ we have

$$\frac{\cot \pi z}{z(z+1)}=\frac{1}{z}\left(\frac{1}{\pi z}-\frac{\pi z}{3}-\frac{\pi^3 z^3}{45}-\cdots\right)\left(1-z+z^2-z^3+\cdots\right)$$
$$=\frac{1}{\pi z^2}-\frac{1}{\pi z}+\cdots$$

which shows that $z=0$ is a double pole with residue $-1/\pi$.
At $z=-1$ we have, putting $t=z+1$,

$$\frac{\cot \pi z}{z(z+1)}=\frac{\cot \pi(t-1)}{(t-1)t}=\frac{\cot \pi t}{(t-1)t}$$
$$=-\frac{1}{t}\left(\frac{1}{\pi t}-\frac{\pi t}{3}-\frac{\pi^3 t^3}{45}-\cdots\right)\left(1+t+t^2+t^3+\cdots\right)$$
$$=-\frac{1}{\pi t^2}-\frac{1}{\pi t}+\cdots$$

which shows that $z=-1$ is also a double pole with residue $-1/\pi$.
Therefore by the residue theorem we have

$$\int_\gamma \frac{\cot \pi z}{z(z+1)}\,dz=\frac{1}{\pi}\sum_1^{N}\frac{1}{n(n+1)}+\frac{1}{\pi}\sum_1^{N-1}\frac{1}{n(n+1)}-\frac{2}{\pi},$$

which, on letting $N \to \infty$, gives

$$0 = \frac{2}{\pi} \sum_{1}^{\infty} \frac{1}{n(n+1)} - \frac{2}{\pi},$$

and hence

$$\sum_{1}^{\infty} \frac{1}{n(n+1)} = 1.$$

8 Fundamental theorem of algebra

1. If $f(z) = z^3$, $g(z) = 9z^2 + 9z + 9$, then for all $|z| = 10$ we have

$$|g(z)| = |9z^2 + 9z + 9| \leq 9|z|^2 + 9|z| + 9 = 999 < 1000 = |z^3| = |f(z)|.$$

Therefore by Rouché's theorem $f(z) = z^3$, $f(z)+g(z) = z^3+9z^2+9z+9$ have the same number of zeros inside $|z| = 10$. But $f(z) = z^3$ has 3 zeros inside $|z| = 10$, all at $z = 0$. Hence also $z^3 + 9z^2 + 9z + 9$ has 3 zeros inside $|z| = 10$, as required.

2. If $f(z) = 9z^2$, $g(z) = z^3 + 9z + 9$, then for all $|z| = 2$ we have

$$|g(z)| = |z^3 + 9z + 9| \leq |z|^3 + 9|z| + 9 = 35 < 36 = |9z^2| = |f(z)|.$$

Therefore by Rouché's theorem $f(z) = 9z^2$, $f(z)+g(z) = z^3+9z^2+9z+9$ have the same number of zeros inside $|z| = 2$, namely 2, since $f(z) = 9z^2$ has 2 zeros inside $|z| = 2$, both at $z = 0$.

3. If $f(z) \equiv 9$, $g(z) = z^3 + 9z^2 + 9z$, then for all $|z| = 1/2$ we have

$$|g(z)| = |z^3 + 9z^2 + 9z| \leq |z|^3 + 9|z|^2 + 9|z| = 55/8 < 9 = |f(z)|.$$

Therefore by Rouché's theorem $f(z) \equiv 9$, $f(z)+g(z) = z^3+9z^2+9z+9$ have the same number of zeros inside $|z| = 1/2$, namely none, since $f(z) \equiv 9$ has no zeros inside $|z| = 1/2$.

4. On $|z| = 1$ we have

$$|z^3 + 6z| \leq |z|^2 + 6|z| = 7 < 8.$$

Therefore $z^3 + 6z + 8$ has no zeros inside $|z| = 1$.

On $|z| = 3$ we have

$$|6z + 8| \leq 6|z| + 8 = 26 < 27 = |z^3|.$$

Therefore $z^3 + 6z + 8$ has 3 zeros inside $|z| = 3$.

5. Take $f(z) = z^4 + 1$, $g(z) = z$. The zeros of $f(z)$ are at ω, ω^3, ω^5, ω^7 where $\omega = e^{i\pi/4}$. Let $\gamma = \gamma_1 + \gamma_2 + \gamma_3$, where γ_1 is the straight line $z = x$ $(0 \leq x \leq R)$, γ_2 is the arc $z = e^{it}$ $(0 \leq t \leq \pi/2)$, and γ_3 is the straight line $z = iy$ $(R \geq y \geq 0)$.

On γ_1 We have $x^4 + 1 > x$. (Clearly!)

On γ_3 We have $|f(z)| = y^4 + 1 > y = |g(z)|$.

On γ_2 We have $|f(z)| = |z^4 + 1| \geq R^4 - 1 > R = |g(z)|$ if $R > 2$.

Hence $|f(z)| > |g(z)|$ on γ if $R > 2$. Therefore $f(z) = z^4 + 1$, $f(z) + g(z) = z^4 + z + 1$ have the same number of zeros inside γ if $R > 2$, namely 1. Argue similarly for the other quadrants.

Appendix 1: Cauchy's theorem

If γ is a closed contour, and if $f(z)$ is differentiable inside γ and on γ, then

$$\int_\gamma f(z)\,dz = 0.$$

Proof
Case 1: γ = Unit square Writing $\gamma = \gamma_1+\gamma_2+\gamma_3+\gamma_4$ where $\gamma_1, \gamma_2, \gamma_3, \gamma_4$ are the four sides taken in anti-clockwise order starting from 0, and writing $f(x+iy) = u(x, y) + iv(x, y)$, we have

$$\int_{\gamma_1} f(z)\,dz = \int_0^1 (u(x,0) + iv(x,0))\,dx,$$

$$\int_{\gamma_2} f(z)\,dz = i\int_0^1 (u(1,y) + iv(1,y))\,dy,$$

$$\int_{\gamma_3} f(z)\,dz = -\int_0^1 (u(x,1) + iv(x,1))\,dx,$$

$$\int_{\gamma_4} f(z)\,dz = -i\int_0^1 (u(0,y) + iv(0,y))\,dy.$$

Therefore,

$$\begin{aligned}
&\int_{\gamma_1} f(z)\,dz + \int_{\gamma_3} f(z)\,dz \\
&\quad = \int_0^1 (u(x,0) - u(x,1))\,dx + i\int_0^1 (v(x,0) - v(x,1))\,dx \\
&\quad = -\int_0^1\int_0^1 \frac{\partial u}{\partial y}(x,y)\,dx\,dy - i\int_0^1\int_0^1 \frac{\partial v}{\partial y}(x,y)\,dx\,dy.
\end{aligned}$$

Also

$$\begin{aligned}\int_{\gamma_2} f(z)\,dz + \int_{\gamma_4} f(z)\,dz &= \int_0^1 (v(0,y) - v(1,y))\,dy + i\int_0^1 (u(1,y) - u(0,y))\,dy \\ &= -\int_0^1\int_0^1 \frac{\partial v}{\partial x}(x,y)\,dx\,dy + i\int_0^1\int_0^1 \frac{\partial u}{\partial x}(x,y)\,dx\,dy.\end{aligned}$$

But the Cauchy–Riemann equations

$$\frac{\partial u}{\partial x} = \frac{\partial v}{\partial y}, \qquad \frac{\partial u}{\partial y} = -\frac{\partial v}{\partial x}$$

hold everywhere inside and on γ. Hence we have

$$\int_{\gamma} f(z)\,dz = \int_{\gamma_1} f(z)\,dz + \int_{\gamma_2} f(z)\,dz + \int_{\gamma_3} f(z)\,dz + \int_{\gamma_4} f(z)\,dz = 0.$$

Case 2: γ = Any rectangle Similar.

Case 3: γ = Any rectilinear contour Meaning $\gamma = \sum_1^N \gamma_n$ where each γ_n is a straight line parallel either to the real axis or to the imaginary axis.

By adding and subtracting further straight lines we can write $\gamma = \sum_1^{N'} \gamma_n'$ where γ_n' are all rectangles. Therefore,

$$\int_{\gamma} f(z)\,dz = \sum_1^{N'} \int_{\gamma_n'} f(z)\,dz = 0.$$

Case 4: γ = Any closed contour We can choose a sequence of rectilinear contours γ_n, all lying inside γ, such that

$$\int_{\gamma_n} f(z)\,dz \to \int_{\gamma} f(z)\,dz$$

as $n \to \infty$.

Appendix 2: Half residue theorem

If γ_r is the contour $z = re^{it}$ $(0 \leq t \leq \pi)$, and if $f(z)$ has a simple pole at $z = 0$ with residue A, then

$$\int_{\gamma_r} f(z)\,dz \to \pi i A$$

as $r \to 0$.

Proof The Laurent expansion of $f(z)$ at $z = 0$ must take the form

$$f(z) = \frac{A}{z} + \sum_{0}^{\infty} a_n z^n.$$

Therefore we must have

$$\int_{\gamma_r} f(z)\,dz = \int_{\gamma_r} \frac{A}{z}\,dz + \int_{\gamma_r} \left(\sum_{0}^{\infty} a_n z^n \right) dz.$$

The first integral evaluates to

$$\int_{\gamma_r} \frac{A}{z}\,dz = \int_{0}^{\pi} \frac{A}{re^{it}}\, ire^{it}\,dt = iA \int_{0}^{\pi} dt = \pi i A.$$

The second integral $\to 0$ as $r \to 0$, since for all $z \in \gamma_r$

$$\left| \sum_{0}^{\infty} a_n z^n \right| \leq \sum_{0}^{\infty} |a_n z^n| = \sum_{0}^{\infty} |a_n| r^n,$$

and therefore by the estimate lemma

$$\left| \int_{\gamma_r} \left(\sum_0^\infty a_n z^n \right) dz \right| \leq \pi r \sum_0^\infty |a_n| r^n$$

which $\to 0$ as $r \to 0$. The result follows.

N.B. We can allow multiple poles provided there are only *odd* negative powers in the Laurent expansion, since these make no contribution to the integral round the small semicircle.

Bibliography

Ablowitz, M.J. and Athanassios, S.F., *Complex Variables*, Cambridge University Press, Cambridge, 1997.
Ahlfors, L.V., *Complex Analysis*, McGraw Hill, New York, 1953.
Copson, E.T., *Theory of Functions*, Oxford University Press, Oxford, 1935.
Knopp, K., *Theory of Functions*, Dover, New York, 1945.
Milewski, E.G., *Complex Variables Problem Solver*, REA, New York, 1987.
Spiegel, M.R., *Complex Variables*, Schaum, New York, 1964.
Titchmarsh, E.C., *Theory of Functions*, Oxford University Press, Oxford, 1932.
Whittaker, E.T. and Watson, G.N., *Modern Analysis*, Cambridge University Press, Cambridge, 1902.

Index of symbols and abbreviations

General index

WITHDRAWN